"Professor Dershowitz is one of the mo[st] [...] our time. He is also renowned and re[...] perspectives. This book's advocacy for vaccine mandates is essential reading for all, regardless of your political leanings."

—Dean Hashimoto, MD, oversees the Workplace Health and Wellness division at Mass General Brigham and is the author of *The Case for Masks*.

"In an era in which many arguments sound like 'I'm right, you're evil,' legal scholar Alan Dershowitz makes his case without demonizing or dehumanizing those with whom he disagrees. Regardless of what you think about government vaccine mandates, this is the only way to engage in a truly liberal conversation."

—Pamela Paresky, PhD, author of *A Year of Kindness*

"The advent of the coronavirus pandemic has faced our society with more difficult questions, and more controversy, than arose during earlier pandemics and epidemics. Opposition to vaccination has arisen within political, religious, and even scientific circles. Alan Dershowitz has tapped into his wide experience as a constitutional scholar and public advocate in order to analyze the issues fairly, often ingeniously. He comes down on the side of compulsory vaccination, (with exceptions as a last resort), but he does so in a manner that should convince even the most diehard skeptic."

—Harvey A. Silverglate, civil liberties and criminal defense lawyer and author of *Three Felonies a Day: How the Feds Target the Innocent*

"Vaccines protect the individual and the community from infectious diseases. Vaccines can decrease transmission in the community and prevent the overburdening of the healthcare systems (e.g., taking up ICU beds). Vaccine mandates have played a major role in minimizing the burden of many infectious diseases. This book provides strong support for vaccine mandates in protecting communities, particularly from a legal point of view."

—Kathryn Edwards, Walter Orenstein, David Stephens, authors of *The Covid-19 Vaccine Guide*

THE CASE FOR VACCINE MANDATES

Also by Alan Dershowitz

What Israel Meant to Me: By 80 Prominent Writers, Performers, Scholars, Politicians and Journalists

Rights From Wrongs: A Secular Theory of the Origins of Rights

America on Trial: Inside the Legal Battles That Transformed Our Nation

The Case for Peace: How the Arab-Israeli Conflict Can Be Resolved

The Case for Israel

America Declares Independence

Why Terrorism Works: Understanding the Threat, Responding to the Challenge

Shouting Fire: Civil Liberties in a Turbulent Age

Letters to a Young Lawyer

Supreme Injustice: How the High Court Hijacked Election 2000

Genesis of Justice: Ten Stories of Biblical Injustice that Led to the Ten Commandments and Modern Law

Just Revenge

Sexual McCarthyism: Clinton, Starr, and the Emerging Constitutional Crisis

The Vanishing American Jew: In Search of Jewish Identity for the Next Century

Reasonable Doubts: The Criminal Justice System and the O.J. Simpson Case

The Abuse Excuse: And Other Cop-Outs, Stories and Evasions of Responsibility

The Advocate's Devil

Contrary to Popular Opinion

Chutzpah

Taking Liberties: A Decade of Hard Cases, Bad Laws, and Bum Raps

Reversal of Fortune: Inside the Von Bülow Case

The Best Defense

Criminal Law: Theory and Process (with Joseph Goldstein and Richard Schwartz)

Psychoanalysis, Psychiatry, and Law (with Joseph Goldstein and Jay Katz)

THE CASE FOR VACCINE MANDATES

ALAN DERSHOWITZ

HOT BOOKS

10 9 8 7 6 5 4 3 2 1

Library of Congress Cataloging-in-Publication Data is available on file.

ISBN: 978-1-5107-7102-4
eBook: 978-1-5107-7104-8

Cover design by Brian Peterson

Printed in the United States of America

Acknowledgments

Thanks to Carolyn, my wife of thirty-five years,
for her inspiration and constructive criticism.
To my assistant Maura for producing the manuscript
from my awful handwriting.

Portions of this book were reviewed and lovingly critiqued
by my medical school grandchildren Lori and Lyle, who
suggested correction but bear no responsibility for any errors.

Dedication

Dedicated to the frontline workers who
risked their lives to save others.

Contents

———

Introduction:
The Libertarian Case
for Vaccine Mandates

A. A Libertarian Approach Derived from
John Stuart Mill

As a lifelong civil libertarian, I generally oppose the government telling individuals what they can and cannot do with
and to their own bodies. That is why I have always favored
a woman's right to choose abortion, a man and woman's
right to have sex and marry anyone they choose, and every
adult's right to refuse medical treatment that will help only
that person.[1] My views derive from those of John Stuart Mill
who brilliantly set out the formula that most civil libertarians have followed for the last century and a half.

> That principle is, that the sole end for which man-
> kind are warranted, individually or collectively, in
> interfering with the liberty of action of any of their
> number, is self-protection. That the only purpose for
> which power can be rightfully exercised over any
> member of a civilized community, against his will is
> to prevent harm to others. His own good, physical or

moral, is not a sufficient warrant. He cannot rightfully be compelled to do or forbear because it will be
better for him to do so, because it will make him happier, because, in the opinions of others, to do so
would be wise, or even right. These are good reasons
for remonstrating with him, or reasoning with him,
or persuading him, or entreating him, but not for
compelling him, or visiting him with any evil in case
he do otherwise. To justify that, the conduct from
which it is desired to deter him, must be calculated to
produce evil to someone else. The only part of the
conduct of anyone, for which he is amenable to society, is that which concerns others. In the part which
merely concerns himself, his independence is, of
right, absolute. Over himself, over his own body and
mind, the individual is sovereign.[2]

A folksier way of putting Mill's doctrine is to say that your
right to swing your fist ends at the tip of my nose.

B. Analysis by Hypotheticals

A useful methodology for determining how this libertarian
approach impacts the COVID vaccine is to begin with hypothetical situations that represented theoretical extremes, and
then to apply them to actual situations that are likely to
occur in the real world. Law professors and philosophers have
employed this "Socratic Method" for generations, as I did
during my 50 years of teaching at Harvard Law School. (It is
less in vogue today, because it is deemed by some to be too
confrontational as a teaching method. I disagree.)

So, here are the two polar extreme hypotheticals. The first posits a vaccine that cures cancer with 100 percent certainty and with no risks or side effects, I would urge everybody to take it. I would want the government to make it available free. I would support incentives to encourage such medical treatment. I might even limit insurance and other benefits to those who refuse to take it. But I would not allow the government to compel any competent adult to take a vaccine that prevents a non-contagious disease from killing only individuals who decline to take it. They have the right to make decisions—even foolish ones—regarding their own bodies, lives, and health.[3] As I put it in the context of smoking cigarettes: everyone has the right to inhale into their own lungs, but not to exhale into mine.

The second hypothetical is imagining a risk-free vaccine that in addition to helping the individual who received it, was also 100 percent effective in preventing the spread of a highly contagious and deadly disease to others (even those who were vaccinated and took additional precautions). I would support a governmental decision, arrived at democratically, that required everyone (with limited medical exceptions) to be vaccinated.

To take this second hypothetical a step further, what if there were a vaccine that did nothing to help the person receiving it, but was 100 percent effective in preventing the spread to others? I think I would favor compulsion, as long as the vaccine was relatively risk free and the disease was fatal or extremely dangerous to others. The human papillomavirus (HPV) vaccine, which prevents serious disease in women, but less so in men, raises the question whether men should be compelled or pressured to take it, because a vaccinated male sex partner reduces the risk to a vaccinated female sex partner.[4]

The above hypothetical situations represent theoretical paradigms that never actually exist in the real world. They are designed to set out the parameters of a real debate about the actual situations we face in the messy world where perfection and absolutes are merely aspirations—like the constitutional goal of forming "a more perfect union." The only thing perfect in our inherently flawed world is the perfect fool who fails to understand that the perfect is the enemy of the good, even in "The best of all possible worlds," as Voltaire's Dr. Pangloss described his pollyannaishly fictional universe. The world of medicine, and especially when it involves new and ever-changing variants of a highly contagious and sometimes lethal virus, is never going to be perfect. It will always require decisions balancing safety against liberty to be made on imperfect, uncertain, and constantly improving information, research, and conclusions.

C. Applying Mill to Real-World COVID

How then would Mill's doctrine—or any other sensible philosophical doctrine—apply to the impact of COVID, vaccines, and other responses to contagious diseases that require compromises with individual autonomy, based on the kind of information we now have, and are likely to get? First it is important to remember that our Constitution does not follow any particular school of philosophy.

As Justice Oliver Wendel Holmes once observed: "The 14th Amendment does not enact Mr. Herbert Spencer's social statics." (Spencer was an influential political philosopher of that age who espoused social Darwinism, in which Holmes himself believed.) Nor does the Constitution enact the principles of John Stuart Mill (in which I believe). But our Constitution does reflect the need to balance appropriately the police powers of the state against the rights of the individual, and Mill's principles have informed judicial, legislative and executive efforts to strike that balance.

With these principles—civil libertarian, Constitutional, and commonsensical—in mind, this short book will consider how to apply them to the current debate about COVID vaccine and other related mandates.

We must begin with the recognition that current COVID vaccines are, unlike my two hypotheticals, far from perfect either in preventing illness or death among those vaccinated (though it is quite effective in that regard) or in preventing its spread to others (though it seems less directly effective in that regard, but indirectly effective by reducing the number and viral load of those infected). Moreover, our knowledge regarding the varying strains of COVID is not only imperfect, it is constantly changing, as the disease itself morphs and poses new challenges to scientists, doctors, philosophers, judges, and citizens. In Chapter 1, I summarize the current research on the relevant issues.

In law as in life itself, we are almost always playing catch up. Government officials, and those who advise them, must make difficult, often life and death, decisions based on probabilities and uncertainties. And research has shown that most people, even very smart ones, are not very good at making probabilistic decisions based on uncertain and everchanging realities.[5] False positives and false negatives often plague efforts to make uncertain predictions. As Yogi Berra aptly put it: "Predictions are hard to make, especially about the future." But they are also necessary to make when we confront an unpredictable pandemic and unpredictable responses to it. It is also necessary to decide whether in given context false positives are preferable to false negatives or vice versa.[6]

D. Calibrated Steps

In the context of COVID vaccinations, governments can take, and have taken, several calibrated steps along the road from voluntary to compelled vaccinations.[7]

The first step, which most thoughtful people would not oppose, is to make the vaccine as widely available and as inexpensive as possible, so that everybody who chooses to take it can have access to its benefits. It should also make widely accessible the most accurate information about the benefits and risks of the vaccine, along with honest disclaimers about the uncertainty of the available information.[8] Almost every country in the world today has taken, or has made efforts to take, this important first step. Because it is largely a question of resources, it has still not been achieved in some parts of the world, and even in some parts of the United States, but it is not particularly controversial and should be encouraged.

The second step, which is controversial, is not to actually compel the vaccine, but to condition access to certain venues on proof of vaccination. This is, of course, a form or degree of de facto compulsion, because access to some of these venues are essential to other rights and needs, such as travel, employment, education, worship, etc. But it still accords the objector the right to maintain his bodily integrity, even if that right is conditioned and costly. It is likely that the courts will uphold reasonable rules regulating access to some venues based on vaccination status.[9]

Closely related to this step is the need for documentary proof of vaccination—which some call "vaccination passports." If access to certain venues is conditioned on vaccination, it follows that proof of vaccination—some documentation—may be required. Some critics compare such documentation to "show me your papers," which they say epitomized repressive regimes. A few absurdly compare it to Nazi Germany and the Holocaust.[10] But post 9/11, documentation has been required by virtually all countries for travel, entering many buildings, voting and other aspects of daily life. I believe the courts will uphold requirements for documentary proof of vaccination to venues that are permitted to condition access on vaccination. Some on the right oppose documentation for vaccination but would require it for voting, while some on the left oppose it for voting but would require it for vaccination. Hypocrisy and inconsistency abound in our toxically divisive culture.

The third and also somewhat controversial step is to compel certain individuals to be vaccinated. These may include members of the armed forces, prisoners, and others. If there were a mandatory draft, this would mean that those drafted would have no choice, just as prisoners have no choice. Today, of course, there is no draft, so a person in the armed forces has the option of not joining or leaving in order to avoid mandatory vaccination.

There is historic precedent for requiring members of the armed forces to be inoculated. Indeed, the precedent goes back to the American Revolution when General George Washington required all of his troops to be inoculated against smallpox, despite the reality that the inoculation available in the 18th century was primitive, not totally effective, and considerably more dangerous than current vaccines. This is what George Washington wrote to John Hancock during the Revolutionary War:

> Finding the smallpox to be spreading much and fearing that no precaution can prevent it from running through the whole of our army, I have determined that troops shall be inoculated. This expedient may be attended with some inconveniences and some disadvantages, but yet I trust in its consequences will have the happiest effects. Necessity not only authorizes but seems to require the measure, for should the disorder infect the army in the natural way and rage with its virulence we should have more to dread from it than from the sword of the enemy.

The rules have always been different for armed forces, and the Washington letter cannot serve as a precedent for civilians though it is informative on the role of necessity in rulemaking.

The fourth and most controversial step would be for the government to mandate vaccination for all citizens who are medically able to be vaccinated without undue risk. It might also exempt religious objectors, though the Constitution probably does not compel such an exemption. There is no direct Supreme Court precedent on the constitutionality of compelled vaccination against COVID, but a 1905 decision authorized a fine for those who refused to be vaccinated against smallpox.[11] Much has changed, both constitutionally and medically, since 1905. Moreover, COVID and its variants are different than smallpox, polio, and other past epidemics. I will explain why it is likely, however, that the courts would uphold the power of government to compel vaccinations (with some exceptions) when deemed necessary to stem the spread of a highly contagious and often lethal virus. Current court cases authorizing restrictions on unvaccinated students, employees, and others point in that direction,[12] but much will depend on the scientific data at the time the cases come before the courts.

I will be discussing each of these calibrated steps in subsequent pages. Because any intrusion on the bodily autonomy of an individual is a compromise with his or her absolute freedom of bodily integrity, I will argue that the government should try the least intrusive measures first and only get to the most intrusive measures if absolutely necessary. Though I do not accept the oft-quoted mantra that "necessity knows no law," I do agree that the law often does, and should take the needs of the day into account in interpreting constitutional and statutory provisions. As Justice Robert Jackson once aptly put it: "The Constitution is not a suicide pact." But neither is it an invitation to denial of liberty. Experience has taught us that governments—even democracies—expand and abuse emergency powers and maintain them even after the emergency has abated.[13] We must be on guard against this phenomenon. In the words of Benjamin Franklin: "Those who would give up essential liberty to purchase a little temporary safety, deserve neither liberty nor safety."

E. In a Democracy, Who Decides?

One of the major questions in any democracy is which institutions or officials get to make the critical, controversial, and often nuanced decisions that require a balance between the safety of society and the liberty of the individual.

Among the components of any jurisprudence are the mechanisms for making the kinds of balancing decisions that will inevitably be required; the checks and balances in any such mechanisms; a recognition that errors will inevitably be made; principles for weighing the costs of different types of errors in different contexts (false positives, false negatives); methods for allocating burdens of going forward and burdens of proof; default rules in situations of equipoise; consideration of absolute (or relatively absolute) prohibitions on, or requirements of certain actions; efforts to integrate new jurisprudential rules into the traditional jurisprudence; processes for evaluation, reevaluation, and change of rules; theories of human action, of sanctions and of the value of life, security, and liberty; the inevitability of unintended consequences, unanticipated events, and unknown elements.

Under our constitutional system of separation of powers and checks and balances, decisions are allocated among the three branches of government. That is true as a matter of constitutional law. But as a matter of public policy, the allocation of responsibility must include medical profession-als, scientific experts, and others whose input should inform the three branches of government. Moreover, administra-tive agencies have largely assumed the role of a fourth branch since the New Deal. I will discuss these issues of allocation of authority as well in the pages to come.

The COVID crisis came upon us so suddenly and unpre-dictably that many important decisions, generally allocated to the legislative branch, were made, on an emergency basis, by the executive branch—by the President, governors, may-ors, administrators, police chiefs, and others whose usual role is to enforce the law made by the legislative branch. But legislative action is far slower and more cumbersome than executive action, which can be taken quickly by a single official. The mandatory closing of businesses, places of wor-ship, entertainment centers, and other venues were ordered by executive, not direct legislative authority. Masking and distancing rules were put in place by governors and mayors. Rent, eviction, and loan moratoria were mandated and extended without explicit legislative authorization. So, too, with some rules regarding vaccinations.

In some instances, these actions had been implicitly del-egated to these executives by prior legislation. Sometimes the executives simply issued orders without express legisla-tive authorization. Sometimes they acted in defiance of con-stitutional limitations. For example, President Biden extended the eviction moratorium without legislative autho-rization with the express knowledge that his action might later be ruled unconstitutional, because he deemed it neces-sary on humanitarian grounds.[14] The Supreme Court has suggested that rent moratoriums—whether authorized by executive or legislative rules—may face constitutional challenges.[15]

In some instances, the legislatures have delegated broad authority to executives to take needed actions during sudden emergencies. In other instances, executives acted on their own without such explicit delegation, claiming "inherent" executive authority during emergencies.[16] These claims are currently being challenged in courts.

On September 9, 2021, President Biden ordered roughly two-thirds of the American workforce to be vaccinated as a condition of their employment. He acted through a series of executive orders and federal rules, but without express legislative authority to punish those (with some exceptions) who do not get vaccinated within seventy-five days. This "emergency" executive action will surely be challenged in court.

F. A Non-Partisan Approach

My analysis in this book will be based on my civil liberties and constitutional perspectives. It will be non-partisan, in the sense that it is not calculated to reflect the views of any political party or faction. It is a shameful reflection of our dangerously divisive age, that even a pandemic and the responses to it divide us along partisan and ideological lines: in general—and with many exceptions—Republicans and conservatives are more skeptical than Democrats and liberals about vaccines, masks, distancing, and other such measures.[17] Extremists on both sides of the political spectrum are also more skeptical than centrists. There are racial, religious, and ethnic divisions as well.[18]

Under my non-partisan, civil liberties, and constitutional perspective, much turns on facts and science: if a vaccine significantly reduces the threat of spreading a serious and potentially deadly disease without significant risks to those taking the vaccine, the case for governmental compulsion grows stronger. If a vaccine only reduces the risk and seriousness of COVID to the vaccinated person but does little to prevent the spread or seriousness to others, the case is weaker. Because we are inevitably dealing in matters of degree—what constitutes "significantly" in the context of reducing or creating risks?—judgment calls will be necessary. Ben Franklin was correct when he talked about not sacrificing "essential liberty" for "a little temporary safety." But what about sacrificing non-essential liberties for a lot of safety? The question of who ultimately gets to make these calls, and who decides what criteria should be employed in making them, are central to any democracy based on the rule of law.

These judgment calls will not be value free. They will inevitably involve—whether consciously or unconsciously—ideological, political, religious, and other biases. Decisions will have to be made about how to act in the face of uncertainty: will there be a presumption in favor of safety or of liberty when the two come into conflict? History suggests that the courts will show considerable deference to legislative and executive decisions that prioritize safety over liberty during a real emergency. I tried to summarize our historical experience in a study I did back in 1971 that focused on executive suspensions of constitutional rights during emergencies:

What then could we reasonably expect from our courts if any American president during a period of dire emergency were once again to suspend important constitutional safeguards? Our past experiences suggest the following outline: The courts—especially the Supreme Court—will generally not interfere with the executive's handling of a genuine emergency while it still exists. They will employ every technique of judicial avoidance at their disposal to postpone decision until the crisis has passed. (Indeed, though thousands of persons have been unlawfully confined during our various periods of declared emergency, I am aware of no case where the Supreme Court has ever actually ordered anyone's release while the emergency was still in existence.) The likely exceptions to this rule of judicial postponement will be cases of clear abuse where no real emergency can be said to exist, and cases in which delay would result in irrevocable loss of rights such as those involving the death penalty. Once the emergency has passed, the courts will generally not approve further punishment; they will order the release of all those sentenced to imprisonment or death in violation of ordinary constitutional safeguards. But they will not entertain damage suits for illegal confinement ordered during the course of the emergency.

But our historical experience—even when tempered by recent developments such as 9/11 and our response to it—ought to teach us that we cannot place our entire reliance upon judges to vindicate our liberties in the midst of a national crisis. Learned Hand recognized this when he said: "Liberty lies in the hearts of men and women; when it dies there, no constitution, no law, no court can save it." But just how deeply is liberty ingrained in the hearts of American men and women? Can we rely on their "eternal vigilance" to resist suspension of fundamental safeguards during periods of crisis? Our historical experience in this respect is disappointing.

G. Masks, Distancing, and Closures

Vaccine mandates are the central focus of this book, because the stakes are highest when government seeks to intrude into decisions involving bodily autonomy and heath. Policies with regard to vaccination may involve issues of life and death. Wearing masks, social distancing, restrictions on businesses, religious, political and other gatherings also raise important issues of liberty, constitutionality, democratic governance, and public policy. Even if less critical than vaccination, they, too warrant serious considerations and will be discussed in the pages to come.

H. The "Least Worse" Solutions

Just as there are no perfect vaccines, there are no perfect solutions to the array of difficult issues raised by this terrible pandemic.[19] In this book, I strive for the best, or perhaps the least worse solutions to dilemmas for which there may not be any perfect or even very good response. Perhaps we will be able to do better as the evidence improves and as we learn more—if we are willing to admit our past errors.[20]

As Judge Learned Hand wisely admonished: "The spirt of liberty is the spirit that is not too sure it is right." Today's zealots are "too sure" they are always right.

Because of the deep divisiveness of our nation even regarding the pandemic, none of my proposed solutions will be without controversy. Some will generate anger.[21]

Few critics will admit they could be mistaken. My goal is to stimulate reasoned discussion and debate, rather than sound bites, bumper-sticker or knee-jerk responses. Unfortunately, we live in an age of recrimination rather than reflection, cursing not compromise, shouting instead of researching, ideology before science, doctrine not data, identity politics over merit, cancellation replacing consideration, nastiness not nuance, rigidity instead of flexibility. The age of enlightenment promised reason over dogma. As Thomas Jefferson wrote in his final substantive letter before his death:

> May [The Declaration of Independence] be to the world, what I believe it will be, (to some parts sooner to others later, but finally to all,) the signal of arousing men to burst the chains under which monkish ignorance and superstition had persuaded them to bind themselves, and to assume the blessings and security of self-government. That form which we have substituted, restores the free right to the unbound exercise of reason and freedom of opinion. All eyes are opened, or opening, to the rights of man. The general spread of the light of science has already laid open to every view the palpable truth, that the

mass of mankind has not been born with saddles on their backs, nor a favored few booted and spurred, ready to ride them legitimately, by the grace of God. These are grounds of hope for others. For ourselves, let the annual return of the day forever refresh our recollections of these rights, and an undiminished devotion to them." Quoted in Alan Dershowitz, *Finding Jefferson* (2008).

Jefferson's promise has been broken by today's extremists, on both sides, who are intolerant of ideas or information that do not comfortably fit into their pre-existing ideological narrative. These know-nothing zealots are the enemies of reason and enlightenment. They are a throwback to the age of dogma—religious, political, and ideological. They are part of the problem we are experiencing as a result of COVID, not the solution we require from science and progress.

Controlling a raging pandemic requires cooperation, coordination, minds open to new ideas, non-ideological compromises, and a willingness to acknowledge mistakes when new information disproves old formulations. Our current culture is not a cause for optimism in confronting COVID. To paraphrase Benjamin Franklin's dire prophecy about hanging separately,[22] we must join forces in combatting COVID, or we assuredly become victims of it.

Chapter 1
The Case for Compulsion:
From Easy to Hard

Under what circumstances may the government constitutionally compel an adult, or a child, to take actions? The government can compel citizens to pay taxes, to serve in the armed forces, and to educate their children. Can the government compel observers to become "good Samaritans" by calling 9/11 if they see a crime in progress? Can they compel citizens to rescue drunks in a puddle or children in a pool if they can do so without endangering themselves? Can the government require drivers to wear seatbelts or motorcyclists to wear helmets? Can it ban cigarette smoking in private as well as in public? Can it compel an STD test as a condition to obtaining a marriage license? Can it compel a pregnant woman to give birth rather than have an abortion? Can it force a driver to submit to a blood alcohol test? Can it make a child have a blood transfusion over his parent's objection? Can it prevent a competent adult from committing suicide? Can it conduct an autopsy on a murdered person over the objection of his family? Can it require the donation of life-saving organs after death? These and similar questions have been asked since the beginning of recorded history. The answers depend on time, place and circumstances, and philosophy.[1]

In general, our Constitution authorizes government to compel actions reasonably deemed necessary to achieve the proper ends of governance: armed forces, fiscal security, education of children. That principle surely includes preventing pandemics from killing large numbers of people. But our Constitution does not justify all means to achieving legitimate ends. It required a constitutional amendment to authorize income taxes, though they were deemed necessary to fiscal security. It would also require amending the Constitution if it were deemed necessary to ban all weapons in high crime areas plagued by gun violence.[2]

So, what about compelling the wearing of masks and social distancing in public places? Presenting a vaccine "passport" or other such documentation as a condition for entering an airplane, train, bus, school, or other places where COVID could be transmitted? Requiring business, places of worship, or school to close down or severely limit the number of people? Mandating vaccination for everyone, except those with medical excuses?

These are questions that today divide our nation. There are no certain answers, though history and precedent provide some guidelines.

A. Mandatory Masking: An Easy Case
The case for mandatory masking in densely crowded public places is an easy one from a constitutional perspective.

There is evidence that masks help stem the transmitability of COVID. It doesn't much matter how much masks help. So long as there is a rational basis for requiring masks, the courts will uphold this mandate. Satisfying this low evidentiary standard is all that is necessary because compelling a person to wear a mask in certain places is a rather minimal intrusion on freedom. There is no compelling evidence that it is a health hazard to ordinary people. It may be an inconvenience, but "facial freedom" is not a constitutional right. Nor is there a compelling moral, political, or legal argument against being required to inconvenience oneself in the interest of protecting others.

Those who have an ideological objection to wearing masks, or to government compulsion, should be allowed to remain unmasked, but without endangering the rest of us. They should go maskless in private, isolated, or other places, where swinging their fists (or ideology) does not hit the tip of my nose (or lungs).

So, the case for mandatory masking is an easy one constitutionally. I think it is also an easy one from a policy perspective, as long as it is done reasonably based on current scientific data: limited to high-risk places; possible exceptions for wearers whose health might be endangered by masks; and reconsidering the requirement when the risks abate.[3]

B. Conditioning Access on Proof of Vaccination: A Strong, but Not Easy Case

Conditioning access to important places on proof of vaccination is a far more consequential form of government compulsion than requiring masks in such places. It essentially requires vaccination as a precondition to engaging with society: traveling (except by car or foot); conducting business; getting an education; attending worship services or mass protests (the latter two are constitutionally protected rights).

It is true that requiring masks and social distancing can also limit one's right of access to these and other activities and venues, but requiring vaccination—and documentary proof—imposes a far greater burden. Compelling an injection into one's body is much more intrusive than compelling the placing of a mask on one's face during certain activities. It is a softer version of compelled vaccination, but it is on the continuum and close to absolute compulsion.

There are, however, other rights and values at stake. As a vaccinated person I have the right to know the level of risk I am incurring going to certain venues. If an unvaccinated person claims a right to access to these venues, they must recognize the right of the vaccinated person to access these venues without undue risk of catching COVID (even if vaccinated individuals have a far lower risk of hospitalization and death). The government has the legitimate authority to choose the safety of the vaccinated over the ideology of the unvaccinated. This is surely true if we are talking about access to constitutionally unprotected activities, such as attending a sporting event, an opera, or a business conference. But it is also true if proof of vaccination is required for exercising an expressly protected constitutional right, such as public worship, political demonstrations, or voting in person. What if it is a fundamental right not expressly protected by the Constitution, such as travel, education, work, family gatherings? What about large celebrations, such as

Obama's 60th birthday bash on Martha's Vineyard or comparable events by Republicans?

Some argue that if a woman has the right to make government stay off her body in the abortion context, she must also have that right in the vaccination context. But pregnancy, unlike COVID, is not contagious.

The courts seem to be moving toward a nuanced approach to these difficult questions of degree. They have said that in making access decisions, government cannot discriminate against religious observances—it cannot make it easier to go to a supermarket than a church if there are comparable risks. Presumably that is the case with other constitutionally protected activities and venues, such as petitioning government, protesting, and voting. But it can condition attending university in person on being vaccinated—presumably with reasonable exceptions and exemptions.

We are witnessing a judicial work in progress. With new decisions coming with some frequency. I predict that in the end the courts will strike a balance that favors safety over ideology in most situations, so long as the safety concerns are reasonable, and the steps taken to protect them are fair and recognize the special status of constitutionally protected rights. If the courts do uphold access restrictions, they will probably uphold proof requirements and documentation—even so-called "vaccination passports." Nor will this turn us into a repressive tyranny, any more than requiring government-issued identification to fly or enter many buildings has done. Falsely crying wolf—or Nazism—hurts rather than helps liberty.

C. Closing down businesses, rent collection, and other sources of income is costly, but may sometimes be necessary.

The COVID pandemic has imposed considerable costs on businesses, landlords, employees, and the economy in general. Elected officials have taken very different views on these controversial issues. Some have favored more restrictive measures designed to slow down the spread of the virus, even at a greater or short-term cost to the economy. Others have favored "opening up" the economy, even if it presents a greater risk of spreading COVID. Both sides are making cost-benefit calculations but weighing the costs differently and assessing short-term versus possible long-term consequences differently.

The courts are not likely to overrule these very different calculations, so long as they are made reasonably and with sensitivity toward constitutional values. The courts may insist that these policy choices must be made by the legislative rather than executive authority, especially if they are not true emergency measures with a short shelf-life.

In a democracy—or a "republic," as the Framers denominated our polity—policy choices are made by the legislature, subject to checks and balances by the other branches. The separation of powers does not authorize the judicial branch to second-guess cost-benefit or long-term versus short-term decisions, even if a judge thinks they are wrong as a matter of policy. Accordingly, we will continue to live with inconsistency from state to state, city to city, and region to region. Such inconsistency is less than ideal in dealing with a pandemic that does not respect political boundaries. But that is the price we pay for our federalist system.

D. Mandatory Vaccination for All Except the Medically Exempt: A Hard but Perhaps Necessary Case

Justice Oliver Wendell Holmes, Jr., once observed that "hard cases make bad law." That is sometimes correct, especially during times of emergency and crises. The case for universally mandated vaccination (even with narrow exemptions) is a hard one. It should be accepted only as a last resort— only if other less intrusive steps have been tried and have not succeeded in significantly stemming the spread of the pandemic. If the only reasonable additional step is universal (with limited exemption) mandatory vaccination, then I believe the courts will uphold it. I make this statement with the knowledge that predicting the decisions of an ever-changing Supreme Court is a risky move. But as Justice Holmes also said, "The role of the law and presumably the lawyer is to make prophecies of what the courts will do in fact." I have little choice, therefore, as a lawyer to try to make such prophecies, even though the Talmud cautioned that "prophecy ended with the destruction of the second Temple," and anyone who claims prophetic insights is either a fool or a knave.

I pride myself in not allowing my personal preferences to cloud my predictions about judicial decisions. I base them on an objective analysis of the law and the facts, which I will now undertake.

All attempts to predict future Supreme Court decisions must begin with past decisions, because of the important role of "stare decisis" ("To stand on decided case") in American jurisprudence. The only past decision that is directly on point to this issue was decided in 1905 in the case of <u>Jacobson v. Massachusetts</u> that involved a Cambridge ordinance mandating vaccination against smallpox and a fine of $5 (today more like $150) for anyone who refused. Smallpox was very different from COVID, and so were the vaccines. Moreover, a small fine doesn't quite meet the definition of "compulsion" or "mandate." Nevertheless, Jacobson is the closest we have as a precedent for mandated COVID vaccination today and tomorrow. So, we must consider that case in some detail.

The majority opinion was written by the highly respected Justice John Marshall Harlan—the dissenting judge in the notorious case of <u>Plessy v. Ferguson</u> which legitimated segregation. (The first Justice Harlan's grandson, with the same name, became a justice in 1955.) The decision upheld the following Cambridge Massachusetts regulation, which had been authorized by a state statute.

> Whereas, smallpox has been prevalent to some extent in the city of Cambridge, and still continues to increase; and whereas, it is necessary for the speedy extermination of the disease that all persons not protected by vaccination should be vaccinated; and whereas, in the opinion of the board, the public health and safety require the vaccination or revaccination of all the inhabitants of Cambridge; be it ordered, that all the inhabitants of the city who have not been successfully vaccinated since March 1st, 1897, be vaccinated or revaccinated."

Henning Jacobson, a resident of Cambridge said he had experienced a bad reaction from a vaccination "when a child," so he refused to comply and was charged by "a criminal complaint." He was found guilty and sentenced to "pay a fine of $5 and the court ordered that he stand committed until the fine was paid." He appealed his conviction to the highest court in Massachusetts, which affirmed it.

Jacobson then sought review by the United States Supreme Court on the following grounds:

That § 137 of chapter 75 of the Revised Laws of Massachusetts was in derogation of the right secured to the defendant by the preamble to the Constitution of the United States, and tended to subvert and defeat the purposes of the Constitution as declared in its preamble:

That the section referred to was in derogation of the rights secured to the defendant by the 14th Amendment of the Constitution of the United States, and especially of the clauses of that amendment providing that no state shall make or enforce any law abridging the privileged or immunities of citizens of the United States, nor deprive any person of life, liberty, or property without due process of law, nor deny to any person within its jurisdiction the equal protection of the laws; and that said section was opposed to the spirit of the Constitution.

The Supreme Court summarily rejected the arguments based on "preamble" and "spirit" of the Constitution and focused on the 14th Amendment as binding authority.

Justice Harlan began by summarizing the authority of the state under its "police power":

> The authority of the state to enact this statute is to be referred to what is commonly called the police power,—a power which the state did not surrender when becoming a member of the Union under the Constitution. Although this court has refrained from any attempt to define the limits of that power, yet it has distinctly recognized the authority of a state to enact quarantine laws and "health laws of every description"; indeed, all laws that relate to matters completely within its territory and which do not by their necessary operation affect the people of other states. According to settled principles, the police power of a state must be held to embrace, at least, such reasonable regulations established directly by legislative enactment as will protect the public health and the public safety.

Harlan then acknowledged that:

> No rule prescribed by a state, nor any regulation adopted by a local governmental agency acting under the sanction of state legislation, shall contravene the Constitution of the United States, nor infringe any right granted or secured by that instrument. A local

enactment or regulation, even if based on the acknowledged police power of state, must always yield in case of conflict with the exercise by the general government of any power it possesses under the Constitution or with any right which that instrument gives or secures.

The court then addressed the key question:

Whether any right given or secured by the Constitution is invaded by the statute as interpreted by the state court. The defendant insists that his liberty is invaded when the state subjects him to fine or imprisonment for neglecting or refusing to submit to vaccination; that a compulsory vaccination law is unreasonable, arbitrary, and oppressive, and, therefore, hostile to the inherent right of every freeman to care for his own body and health in such way as to him seems best; and that the execution of such a law against one who objects to vaccination, no matter for what reason, is nothing short of an assault upon his person.

The opinion then provided a general answer:

But the liberty secured by the Constitution of the United States to every person within its jurisdiction does not import an absolute right in each person to be, at all time and in all circumstances wholly freed from restraint. There are manifold restraints to which every person is necessarily subject for the common good. On any other basis organized society could not exist with safety to its members. Society based on the rule that each one is a law until himself could soon be confronted with disorder and anarchy. Real liberty for all could not exist under the operation of a principle which recognizes the right of each individual person to use his own, whether in respect of his person or his property, regardless of the injury that may be done to others.

It then applied these generalities to Jacobson's specific appeal:

The legislature of Massachusetts required the inhabitants of a city or town to be vaccinated only when in the opinion of the board of health, that was necessary for the public health or the public safety. The authority to determine for all what ought to be done in such an emergency must have been lodged somewhere or in some body; and surely it was appropriate for the legislature to refer that question, in the first instance, to a board of health composed of persons residing in the locality affected, and appointed, presumably, because of their fitness to determine such questions. To invest such a body with authority over such matters was not an unusual, nor an unreasonable or arbitrary, requirement. Upon the principle of self-defense, of paramount necessity, a community has the right to protect itself against an epidemic of disease which threatens the safety of its members. It is to be observed that when the regulation in question was adopted smallpox, according to the recitals in the regulation adopted by the board of health, was prevalent to some extent in the city of Cambridge, and the disease was increasing. If such was the situation,—and nothing is asserted or appears in the record to the contrary,—if we are to attach any value whatever to the knowledge which, it is safe to affirm, in common to all civilized peoples touching smallpox and the methods most usually employed to eradicate that disease, it cannot be adjudged that the present regulation of the board of health was not necessary in order to protect the public health and secure the public safety. Smallpox being prevalent and increasing at Cambridge, the court would usurp the functions of another branch of government if it adjudged, as a matter of law, the mode adopted under the sanction of the state, to protect the people at large was arbitrary and not justified by the necessities of the case.

The court then addressed an objection that may sound familiar to current ears: That some doctors "attach little or no value to vaccination as a means of preventing the spread of smallpox, or who think that vaccination causes other diseases . . ." Justice Harlan's answer also has relevance to today's debate:

We must assume that when the statute in question as passed, the legislature of Massachusetts was not unaware of these opposing theories, and was compelled, of necessity, to choose between them.

It then quoted with approval from a New York State decision:

The fact that the belief is not universal is not controlling, for there is scarcely any belief that is accepted by everyone. The possibility that the belief may be wrong, and that science may yet show it to be wrong is not conclusive, for the legislature has the right to pass laws which, according to common belief of the people, are adapted to prevent the spread of contagious diseases. In a free country where the government is by the people, through their chosen representatives, practical legislation admits of no other standard of action, for what the people believe is for the common welfare must be accepted as tending to promote the common welfare, whether it does in fact or not. Any other basis would conflict with the spirit of the Constitution and would sanction measures opposed to a Republican form of government. While we do not decide, and cannot decide that vaccination is a preventive of smallpox, we take judicial notice of the fact that this is the common belief of the people of the state, and, with this fact as a foundation, we hold that the statute in question is a health law, enacted in a reasonable and proper exercise of the police power.

Finally, the court appended a caveat to its broad decision:

> Before closing this opinion we deem it appropriate, in order to prevent misapprehension as to our views, to observe—that the police power of a state, whether exercised directly by the legislature or by a local body acting under its authority may be exerted in such circumstances, or by regulations so arbitrary and oppressive in particular cases, as to justify the interference of the courts to prevent wrong an oppression. Extreme cases can be readily suggested. Ordinarily such cases are not safe guides in the administration of the law. It is easy for instance, to suppose the case of an adult who is embraced by the mere words of the act, but yet to subject whom to vaccination in a particular condition of his health or body would be cruel and inhuman in the last degree. We are not to be understood as holding that the statute was intended to be applied to such a case, or, if it was so intended that the judiciary would not be competent to interfere and protect the health and life of the individual concerned.

But the court distinguished such an "extreme" situation from claims that a vaccination may be "distressing, inconvenient, or objectionable to some." In such cases, it is "The duty of the constituted authorities primarily to keep in view the welfare, comfort and safety of the many and not permit the interests of the many to be subordinated to the wishes or convenience of the few."[14]

The court thus upheld the regulation, rejecting Jacobson's challenge.

Much has changed—medically, legally, politically, journalistically—since 1905, but the fundamental approach articulated in the Jacobson decision remains essentially unchanged: the government has the "police power" to compel reasonable intrusions into the body of protesting individuals if such intrusions are necessary to prevent the spread of a highly contagious and oft-times lethal pandemic, so long as it does not exercise that power in an "arbitrary and oppressive [manner] in particular cases."

Today's Supreme Court may use different words and formulas in deciding how to strike the constitutionally proper balance between community safety and individual liberty. It requires different levels of scrutiny—"rational basis," "intermediate," or "strict"—depending on the nature of the intrusion. But in the end, it will probably follow the general approach taken by the Jacobson court. It may require more than a mere "rational" evidentiary basis before it legitimates compelled vaccination. It may demand explicit exemptions based on medical and perhaps religious objections. It may move slowly, waiting for lower courts to sort out the varying issues regarding vaccinations. It may not even have to confront the fundamental issue, because governments may not mandate universally compelled vaccinations if less intrusive steps—or nature itself—succeed in stemming or slowing down the spread of COVID. Only time will tell.

I do not believe that the current Supreme Court will directly overrule the Jacobson precedent, despite the changes. It will probably follow its general guidelines and add some that reflect the current realities and constitutional jurisprudence.

The bottom line is that the Constitution, as interpreted by the current courts, will probably not stand in the way of reasonable masking, distancing, business closing, venue limiting, and vaccination mandates—as long as they don't discriminate against constitutionally protected activities, as long as the available scientific information supports the benefits of such measures, and as long as the mandates have proper legislative authority.

There is little remaining scientific doubt that the available vaccines, perhaps with the need for booster shots for some or all recipients, do benefit those who are vaccinated. Even if they don't eliminate the likelihood of contracting COVID, they definitely reduce, quite significantly, the seriousness of the resulting disease in most vaccinated adults. The data show that vaccinated individuals who catch COVID require less hospitalization and die less frequently.

Unvaccinated adults require more hospitalization, more use of ventilators, and more access to intensive care units. Thus limiting or denying these potentially lifesaving facilities to the vaccinated who suffer from breakthrough COVID.

These results alone might not justify, based on Mill's formulation, mandating vaccination for those who refuse to accept the individual benefits of the vaccine because of its perceived risks, or for other ideological, religious, or political reasons. But beneficent results for individuals, combined with positive results for others, would justify compulsion, both by Mill's formulation and by the Constitution. The empirical question that may determine the constitutional outcome is whether and to what degree compelled vaccination prevents or curbs the spread of COVID to individuals who have opted for being vaccinated. Put another way, do unvaccinated individuals pose a danger to vaccinated individuals? Do the swinging fists of the unvaccinated hit the noses and lungs of the vaccinated?

According to current research, the kind of "herd immunity" that we were able to achieve with other pandemics[5] is unlikely to be achieved for COVID and its variants. Sir Andrew Pollard, the head of the Oxford Vaccine Group, told British Parliamentarians that, "We are in a situation here with this current variant where herd immunity is not a possibility because it still infects vaccinated individuals."[6] Dr. Andrew Fredman, another expert, agreed but added: "However, even without complete 'herd immunity,' the higher the proportion of the population fully immunized, the lower the incidence of infection in the community."[7] This is because research suggests that fully vaccinated people may be somewhat less likely than unvaccinated people to pass on the virus.[8] Yet another expert has opined that the herd immunity "mathematical model could not easily be applied to an 'unprecedented' virus like COVID that was still little understood with diverging, globally circulating variants emerging."[9] But he added nonetheless that "the more people on the globe effectively vaccinate, the fewer viral copies we'll have on the planet, thus the less spread to mutate and spread the next wave of variants."[10]

Other experts are even more optimistic:

Researchers in Israel studied vaccinated people who became infected. The viral load in these break-through cases was about three to four times lower than the viral load among infected people who were unvaccinated. Researchers in the U.K. reported a similar result.

They also found that vaccinated people who became infected tested positive for about one week less than unvaccinated people.

We also now have evidence that infected people with lower viral load spread the virus to fewer people, based on contact-tracing studies in the U.S., India, and Spain. This is supported by laboratory research demonstrating the nasal samples from infected people with lower viral load are less likely to contain infectious virus. . . . The totality of these impressive data should bolster confidence that the COVID vaccines are extraordinarily effective in reducing transmission of the virus. This suggests that vaccinating a large majority of Americans throughout the country is our surest bet for returning to normal. Entirely eliminating spread of the virus may be an unreachable goal, but mass vaccination—in the U.S. and around the world—will relegate COVID to the background of our lives.[11]

None of this is absolutely certain or permanent, as I point out in the next chapter. All of it is subject to change as new information and research become available.[12] Recently, the Centers for Disease Control and Prevention issued studies showing that, "Although the vaccines remain highly effective against hospitalizations and deaths, the bulwark they provide against infection with the virus has weakened in the past few months."[13] There is dispute over whether this warrants booster shots for all.

The research and debate continues. But in the meantime—and when it comes to the science of ever-varying diseases we <u>always</u> live in the meantime—governments are entitled to resolve doubts in favor of the scientific evidence that suggests that mass vaccination—which may require compulsion—has the potential to slow down the spread of the virus and its variants. That is enough to satisfy constitutional requirements.

Nor is it a constitutionally acceptable answer to say that the vaccinated are sufficiently protected by the benefits of vaccination because they are unlikely to need hospitalization or to die. They have the right to be protected even from less lethal manifestations of the disease. Moreover, the current data cannot assure that there will not be long term effects of non-symptomatic or less-symptomatic COVID. To be sure, the current data cannot assure that there will be no long-term effects from being vaccinated, but the Jacobson case suggests that governments have the police power to make judgments weighing the likely costs and benefits—both short and long-term—based on the available, even if less than certain, scientific data.

Governments can, of course, be wrong, as they were grievously wrong when several states authorized mandatory sterilization of "mental defectives" in the early 20th century.

The Supreme Court, in an opinion authored by Justice Oliver Wendell Holmes and joined by Justice Louis Brandeis and six other justices, cited the Jacobson case in support of mandatory sterilization of "epileptics and feeble minded" persons. Here is the court's dangerously overbroad reasoning:

> We have seen more than once that the public welfare may call upon the best citizens for their lives. It would be strange if it could not call upon those who already sap the strength of the State for those lesser sacrifices, often not felt to be such by those concerned, in order to prevent our being swamped with incompetence. It is better for all the world, if instead of waiting to execute degenerate offspring for crime, or let them starve for their imbecility, society can prevent those who are manifestly unfit from continuing their kind. The principle that sustains compulsory vaccination is broad enough to cover cutting the Fallopian tubes. <u>Jacobson v. Massachusetts</u>, 197 U.S. 1, 25 S.Ct. 358, 49 L. Ed. 643, 3 Ann. Cas. 765. Three generations of imbeciles are enough."

This now maligned and overruled decision[14] was cited by Nazi defendants at Nuremberg in defense of their decisions to kill the feebleminded.

The unjustified extension of the "principle that sustains compulsory vaccination" to the very different and far more intrusive "cutting of the Fallopian Tubes," should give every civil libertarian—indeed every decent person—pause. So should the uncritical acceptance of the science of the day by the Supreme Court. It's difficult to imagine a more absurd argument than equating an injection to a permanent deprivation of the right to reproductive freedom. Yet, eight justices accepted this argument over only one dissent. Eugenics—the improvement of humankind by regulating reproduction as in animal breeding—was all the rage at Harvard and other universities in the 1920s. This misapplication of biologic Darwinism to social Darwinism clearly influenced Holmes, Brandeis, and other justices and policy makers. It reminds us that, as I argue in the next chapter, we must be skeptical of scientists, and the abuse of science, while encouraging scientific experimentation and innovation.

The current scientific consensus would seem to support the commonsense conclusion that universal or near universal vaccination will help to prevent or slow down the spread of COVID, even if vaccinated individuals can still be contagious. The precise mechanisms through which this occurs may not yet be known with certainty, but governments have the power to err on the side of safety benefits, when the cost of such safety is nothing more intrusive than an injection, as distinguished from the cutting of Fallopian Tubes which deny the human right to have children. But just because a rightly decided case was used as a precedent for a wrongly decided case does not by itself disqualify the original precedent, though it suggest caution in expanding it. Many "good" precedents have been used to bolster "bad" cases.

Buck v. Bell was wrong on the science, wrong on the morality, and wrong on the law. It should give us pause on any intrusion into bodily integrity, but pausing is different than not acting to try to prevent the spread of a highly contagious and dangerous disease. We have no alternative but to act, even in the absence of complete scientific certainty and unanimity of opinion.

The case for mandatory vaccinations against COVID, as a last resort, with reasonable exemptions, is a hard one. But a solid constitutional case can be made for it, if it becomes necessary, which I surely hope it does not.

Hard cases may make bad law, but the courts cannot avoid making decisions in hard cases, including those pitting the safety of the many against the liberty of the few. All we can ask is that they make wise decisions based on the best available science, and that they remain open to reconsider their decisions as the science changes.

Chapter 2
The Case for Science and Skepticism

On March 18, 2020, at the very beginning of the COVID-19 pandemic, I published an op-ed entitled, "Believe Science, but Be Skeptical of Scientists." I think subsequent events have borne out my skepticism. Here is what I wrote back then.

I am a skeptic by nature. I never believe what I read or hear without independently checking it. So, when I read that public health officials were urging people not to buy face masks, because they don't work, I was doubtful.

The officials also said that if individuals buy face masks in large numbers, there won't be enough for health providers. That I believed. But the combination of reasons—they don't work, but they are important for health providers—immediately set off alarm bells in my skeptical mind.

If they don't work for ordinary individuals, why should they work for health providers?

Maybe there is a relevant difference. I kept an open but skeptical mind, while wearing the single N95 mask that I bought just in case.

It now turns out that the public health officials who were telling us not to buy masks were not telling us the whole truth. They were giving us only half the equation.

While it's true that a mass run on masks might deny them to health providers, it's equally true that masks may provide some layer of protection above and beyond the other precautions that everyone should take, such as handwashing and social distancing.

Those who misled us did so deliberately, but with a benign motive: they truly believed that it was more important for health providers to have masks than for every individual to stock up on them. When providers get sick, it has a greater impact on public health than if ordinary individuals catch the virus.

In order to make sure that individuals did not place their own safety above that of the community, a decision was made to present the facts in a skewed manner to disincentivize private purchases of masks. Although well-intentioned, this deception has apparently backfired.

Many people saw through the ruse and thought that what was good for health providers was good for them and their family, and they stocked up on masks. So we now have a situation where there has been a run on masks, while at the same time there has been a diminution in the credibility accorded those in charge of telling us how to react to the crisis. The worst of both worlds.

Honesty may not always be the best policy in extreme emergencies, but dishonesty—even when positively motivated—is not likely to work for long in a society in which the social media amplifies the voices of critics, and reasonable people don't know who to believe.

Another claim about which I was skeptical was that the virus is only contagious by physical contact with infected individuals or surfaces which they touched.

Over and over again, it was emphasized that this particular virus could not be caught by airborne or aerosol transmission. In other words, it doesn't travel through the air. I was skeptical of this claim because it seemed inconsistent with the speed and frequency with which transmissions were occurring around the world.

I told my friends and family to act as if they could get the virus through the air. There is no downside to being more careful.

Recently research has confirmed my skepticism. It now turns out that the virus can remain suspended in the air for a relatively short time, though it loses its potency while falling to the ground. This means that we are at far greater risk of catching the virus even if we wear gloves, wash our hands, and avoid touching surfaces.

It probably also means that masks may be even more important than we were previously led to believe, even if we were skeptical about the "masks don't work at all" message.

These are only two examples of what are sure to be other false messages we have been receiving in the early stages of the pandemic. As more data emerges, we will receive more advice from scientists, most of which will probably be accurate, but some of which will almost certainly turn out to be less than fully accurate.

How should we assess this mélange of information, mis-information, partial truths, and outright falsehoods to which we are certain to be exposed? It won't be easy, especially in the age of social media, where everyone is an expert, and all opinions are created "equally."

A cartoon that was recently circulated makes the point. It has a typical guy looking at his computer and saying: "That's odd: My Facebook friends who were constitutional scholars just a month ago during President Trump's first impeachment trial are now infectious disease experts. . . ."

Chapter 3
There Is No Religious Right to Refuse Vaccination

Even before the COVID-19 pandemic struck, I wrote about the issue of religious exemptions for mandatory vaccinations in the context of the measles vaccine: What I said about that vaccine is applicable to COVID vaccines.

New York just eliminated all religious exemptions for mandatory measles vaccination. It was the right thing to do. There is no constitutional basis for requiring a religious exemption. Nor, in my view, are there any plausible religious arguments against mandatory vaccinations to spread communicable and potential lethal diseases.

Let's begin with the religious arguments: _____.

I have left this blank because there are none. I read widely in religious literature, especially Jewish literature. I have never come across a coherent religious argument against mandatory vaccination for deadly contagious diseases. Jewish law has an overriding religious concept called "pikuach nefesh," which elevates the protection of human life over virtually every other value.

The Jewish Bible is scrupulous in demanding protecting against communicable diseases such as leprosy. There is nothing in Jewish law that requires the parents to turn their children into "typhoid Marys," infecting friends, family, classmates, and neighbors.

The claimed religious argument is rejected by the vast majority of rabbis of every denomination, including by most ultra-Orthodox and Hasidic rabbis. Only a handful of marginal rabbis preach this anti-Jewish and anti-life philosophy.

I challenge any rabbi to debate me on the Jewish religious law regarding vaccination and communicable diseases. He will lose the debate because there is simply no basis in Jewish law for any such argument. Religion is being used as a cover for a misguided political, ideological, conspiratorial, and personal opposition to vaccination. Don't believe any rabbi who tells you otherwise.

Now we can turn to the constitutional argument: _____.

Another blank, because there is none that would permit parents to refuse to vaccinate a child against a communicable disease, even if there were plausible religious reasons for their decisions (which there are not).

There are three basic categories of compelled medical intervention about which the Constitution has something relevant to say.

The first category involves compelling a competent adult to take lifesaving measures to prevent his own death. There are strong constitutional and civil liberties arguments against such compulsion. It really doesn't matter whether the opposition to such measures is religious or philosophical.

An adult Jehovah's Witness may have a strong First Amendment claim against receiving a blood transfusion to save his or her life. But an atheist would also have a compelling argument. Indeed, Jewish law is more protective of life than American constitutional law: Jewish law prohibits a competent adult from refusing a lifesaving medical procedure. It also prohibits suicide.

The second category is where a parent is being compelled to employ lifesaving medical procedures to save the life of a child. The courts generally require the parent to save the life of the child. So, a Jehovah's Witness child could be compelled to receive a blood transfer without regard to their parent's religious objection.

Now we get to the third category, the one about compelled measles vaccination. A parent does not have a constitutional right to refuse to vaccinate a child against a highly contagious and potentially lethal disease which might kill that child (category 2) but might also kill a friend or neighbor who doesn't share the parent's religious view (category 3).

That is about the easiest constitutional question I have ever confronted. There is no compelling argument against requiring a child to be vaccinated against communicable diseases regardless of the parents' wishes and regardless of whether their objections are religious or secular.

Theoretically, a parent could move their case from category 3 to category 2 (or an adult could move it to category 1) if there were an assurance that they would spend all their lives in a bubble that prevented contagious diseases from spreading. Or perhaps in a community of anti-vaxxers who would spread the disease only to other anti-vaxxers. The difference is between hurting oneself and hurting others. The common analogy is that a smoker in a closed public space might have the right to inhale, but not exhale.

This civil libertarian position goes back to John Stuart Mill and even further in intellectual history. I can think of no thinker in the history of the world who has ever advocated the anti-vaccine position. There is no coherent argument—religious, constitutional, civil libertarian, or commonsensical—in favor of allowing people to refuse to be vaccinated against communicable diseases. So why then did so many state legislators in New York vote against the bill? Politics, not principles.

Chapter 4
A Vaccine Requirement Is Not "Nazi Experimentation"

S hortly after the outbreak of COVID-19, extremists started to argue that compelling a vaccine or conditioning employment on being vaccinated is akin to medical experimentation done by the notorious Nazi Doctor Mengele. I wrote this in June of 2021:

The trend seems to be going in the direction of requiring and approving compulsory vaccinations as a condition of employment. A Texas federal judge recently required a nurse at Houston Methodist Hospital either to be vaccinated or to lose her job. He correctly characterized as "reprehensible" the nurse's argument that a vaccination requirement is akin to medical experimentation done during the Holocaust.

The medical experimentation done by Dr. Josef Mengele and others in Auschwitz was designed to kill the patients, not to help them. Vaccines are designed to save lives. To make any analogy to the Holocaust is to suggest that the Holocaust was no worse than vaccination. That is a form of Holocaust denial, deserving only of condemnation.

But even without that exaggerated, bigoted hyperbole, the argument offered by the Houston nurse and her fellow plaintiffs is not constitutionally sound.

The United States Supreme Court, more than 100 years ago, ruled that the public health power of government extends to mandating vaccines against highly communicable and often lethal diseases. In that case it was smallpox. In this case it is COVID-19. It's up to the government to determine the safety requirements for vaccinations, and here it has been determined that, in the light of the seriousness of the current pandemic, the vaccine is safe enough.

The Texas court went out of its way to emphasize that nobody is threatening the nurse with imprisonment. She has a choice: She can refuse to be inoculated, but she cannot work in the hospital if she makes that decision. That is a perfectly rational judicial conclusion.

The hard case may never come. That would be if all Americans were required to be vaccinated without regard to religious or philosophical beliefs. We haven't reached that point yet because there are still more people who want to be vaccinated and haven't received their doses than there are conscientious objectors. It is unclear whether we will be able to reach herd immunity without some kind of compulsory vaccination, but we can come a lot closer than we now are.

During the Revolutionary War, Gen. George Washington required all of his troops to receive the primitive vaccinations then available to prevent the spread of smallpox. I am aware of no objection to that order, nor to the modern-day mandate that all American military personnel must be vaccinated against multiple diseases.

Even for those who oppose vaccination on medical or ideological grounds, sacrifices are often required as a condition of living in a free and democratic society. Many people don't want to pay taxes, or to send their children to school, or to show ID when they get on airplanes. But the law requires them to do so.

In general, our nation has provided exceptions for conscientious objections, based on religion or closely related philosophical beliefs. But this doesn't mean that conscientious objectors for vaccination should be allowed to endanger others. Recall that the vaccine is only about 95 percent effective. No one would get on an airplane if there was a 5 percent chance of a crash—nor should a vaccinated person be required to encounter an unvaccinated person. So perhaps conscientious objection should be permitted, but it should be conditioned on not exposing others.

The issue of mandatory vaccination is emotionally wrought. Unfortunately, like everything else in America today, it has been caught up in politics. Extremes on both sides of the political spectrum are more opposed to it than people at the center, and people on the center-right seem more skeptical than people on the center-left. Skepticism is healthy in a democracy. As the great jurist Learned Hand once said, "The spirit of liberty is the spirit that is not too sure that it is right."

But, in a democracy, emotional issues are resolved under the rule of law—by legislation, executive orders, and judicial review. That process is now going forward. There will probably be decisions both ways, and they will vary with the facts of each case. In the end, public health considerations almost certainly will prevail over any individual preferences. That is not a sign of tyranny. It is a sign of democracy at work.

Chapter 5
Debating Vaccination with Robert F. Kennedy Jr.

———————

Shortly after the pandemic began, I agreed to debate Robert F. Kennedy Jr. on the issues surrounding COVID vaccines.[1] It ran on July 23, 2020. Much has changed since then, but essential issues persist. Here is a slightly edited transcript of the debate. The debate was moderated by Patrick Bet-David, host of the Valuetainment podcast.[2]

PATRICK BET-DAVID: We've invited a lot of different doctors to want to come and debate the topic of COVID-19 vaccine and everyone's turned it down. . . . A video popped up about what Alan Dershowitz said that led to a dialogue with Robert Kennedy. And then we said, what we can set up a friendly debate here together. They both agreed. . . . So, first thing I wanna do is I wanna share my screen and I want the audience to see what led us here. A comment that you made on a podcast you did. And then I'll go from there asking your thoughts on it. So, here's what was said in an interview a few weeks ago by Alan. Let's show a clip of this:

CLIP OF INTERVIEW:

Dershowitz—Let me put it very clearly. You have no constitutional right to endanger the public and spread the disease even if you disagree. You have no right not to be vaccinated. You have no right not to wear a mask. You have no right to open up your business.

Moderator on video clip—Wait, can I stop you? No right not to vaccinated? Meaning if they decide you have to be vaccinated, we have to be vaccinated?

Dershowitz—Absolutely, and if you refuse to be vaccinated, the state has the power to literally take you to a doctor's office and plunge a needle into your arm.

Moderator on video clip—Where is that in the Constitution?

Dershowitz—If the vaccination is designed to prevent the spreading disease, if the vaccination is only to prevent a disease that you will get, for example, if there's a disease that will kill you, you have the right to refuse that, but you have no right to refuse to be vaccinated against a contagious disease. Public health, the police power of the constitution gives the state the power to compel that. . . .

BET-DAVID: Alan, those are some strong statements you made. Obviously, his reaction, a lot of people's reactions, has your position changed since making those statements on that interview?

DERSHOWITZ: The statements I made on the interview were professional statements based on reading Supreme Court cases, not expressing personal views. I have strong personal views, but my constitutional views haven't changed at all. Let me be very clear. I don't think this issue is going to come up in the near future because right now, the *New York Times* has a big story today in which they talk about how there's gonna be a limited number of vaccines and people are gonna be waiting in line to get them. So, the issue is not gonna be confronted as to mandatory vaccines.

You know, having said that, I wanna just pause for one second and say how important this debate is and how privileged I am to participate in it with a so distinguished conversationalist as Robert Kennedy. I, of course, knew his father. I had actually been offered a job to work with his father when he was Attorney General, but Harvard offered me a job and I decided to take it. I was a great fan of Senator Robert Kennedy, Attorney General Robert Kennedy. And I think I consider myself a friend of the Kennedy family, and I consider myself a friend of Robert Kennedy. I admire enormously the environmental work he's done. And I think he's performed an important function by raising issues about vaccination.

We, as you'll see in this conversation, we'll disagree. We'll probably agree on more things. People will be surprised at our level of agreement, but on the issue of constitutionality, I am confident that this Supreme Court would follow a Supreme Court precedent from 1905 and would say if there is a safe, and that's crucially important, effective measure that could significantly reduce the contagious impact of a deadly disease, like the current pandemic virus, that the state would have the power to either directly compel vaccination. Or, for example, condition young students coming to school on being vaccinated or people doing other things that might result in contagion being vaccinated. So, no, I haven't changed my professional constitutional opinion, but as Robert will tell you, we've had conversations offline, and he has persuaded me about a number of things relating to the health and safety and efficacy of vaccines. So, I've learned a lot from our conversations and I hope people will learn something from our conversations today, but the constitutional issue in my mind remains the same.

BET-DAVID: So, it's important to unpack that constitutionally, you remaining same position to say, if the government wanted to mandate and make us take a vaccine, we can't say anything to it. That position is not changing.

DERSHOWITZ: That's right. As long as the vaccination is safe and effective. An example, if you have somebody who has unique vulnerability to vaccinations, that person might get a medical exemption. The issue of religious exemption is something the courts have considered most recently. The Supreme Court did just in the last day or two create religious exemptions for private schools, religious schools in terms of whether employment laws operate on religious schools. So we would have to see what the court would say about religious exemptions, but as a general matter, a healthy person who simply has an ideological objection to vaccinations as such, not to this particular vaccination, because of health reasons or vulnerability, the Supreme Court would, I predict, hold that the state could in one way or another compel vaccination either directly or as a condition of people engaging in public activities or activities that could create contagion. Yeah, that's my position.

BET-DAVID: That's very important to know because there's your personal beliefs, which is completely different than what you think will be able to be mandated. So, having said that, Robert, I know you've seen this before when this statement was made and in one case, Alan even said he'd be willing to debate Robert Kennedy on this topic, which kind of led to us wanting to do this debate. What was your initial reaction of watching what Alan said and what has changed since you and him have had calls together offline?

KENNEDY: Well, I want to begin by saying thank you Alan for participating in this debate. I've actually been trying to a debate on this issue for 15 years. I've asked Peter Hotez, I've asked Paul Offit, I've asked all of the major leaders who are promoting vaccines to debate me and none of them have. And I think it's really important for our democracy to be able to have spirited, civil discussions about important issues like this. This is an issue that's been on the news 24 hours a day for the last four months, and yet there's no debate happening about this. It's all kind of a repetition of the government orthodoxies and government proclamations and democracy functions only when we have the free flow of information. Policy is best often crafted in the furnace of a heated, spirited debate. It is part of our constitutional system, it's part of America's tradition. We invented free speech in this country. Is it the First Amendment? And it ought to be something that we celebrate and that we model the world. It shouldn't be something where you now have democratic leaders, like Adam Schiff, calling on social media sites to censor debate about an important issue. That shouldn't happen.

So, I'm very grateful that Alan, who I know loves the First Amendment, for actually agreeing to debate on an issue in which he is at a disadvantage because I've spent fifteen years working on this issue. I'm in a big disadvantage to him when it comes to talking about constitutional law and I'm gonna try and keep a lot this debate on my side of the issue.

Let me start out by saying I don't agree with Alan's, and this is a very small disagreement, because Alan and I have talked a lot offline and I think we've come to a place where we really believe this is gonna be a conversation, not a debate, because I think on most of the issues we are in agreement and he made the qualifiers when he came up and he said if it's safe, if it's effective. And I think those are the big ifs at the playground where this debate is really happening. And I think in the end, he and I would end up in the same place in that debate.

I will make a minor dispute, which is the Jacobson case, which was decided in 1905 was not a case where the state was claiming the power to go into somebody's home and plunge a needle into their arm or kick down their door and take them by force. The Jacobson actually what the guy who was resisting was taking the smallpox vaccine. He was from Cambridge, Massachusetts, and the penalty for not taking the vaccine was a $5 fine. It's like a traffic ticket. . . . He had been injured in a previous vaccine, so he didn't want to take this one. He took the case to the Supreme Court, he lost and the remedy was he paid a $5 fine.

There's a big constitutional chasm between that remedy, which is paying a fine, and actually going in and holding somebody down and forcibly injecting them. And I'm not convinced the Supreme Court of the United States at this point would uphold that kind of law, nine to zero or eight to one at all. So, let me just say that. Let me now go to the initial place where I think we're in agreement. I think Alan and I are both in agreement that this should be a voluntary program.

If there's mandates, it should be as an ultimate, final, dramatic drastic remedy. And the question is, why can't we do a voluntary program? When Alan and I were kids people wanted to get vaccinated. There was no fear of this statement of the polio vaccine and people had a tremendous trust in our health regulatory officials. And today, that trust has evaporated to the extent were now 50 percent of the people who are polled in this country are saying they may not take the COVID vaccine and 27 percent are hard no. This is even before the vaccine is developed. Why is that happening? That's the question I think we really have to ask ourselves.

DERSHOWITZ: I agree.

KENNEDY: Why do so many Americans no longer trust our regulatory officials and trust this process? And one of the reasons is, vaccines are very, very interesting and very different kind of medical prerogative because it's a remedy that is being. It's a medical intervention that is being given to perfectly healthy people to prevent somebody else from getting sick. And it's the only medicine that is given to healthy people. So, you would want, and particularly to children who have a whole lifetime in front of them, so you would expect that we would want that particular intervention to have particularly great risk guarantees that it's safe 'cause you're saying to somebody, we are going to make you make this sacrifice for the greater good. You have no health problems; you have zero risk of this disease. We are gonna force you to undergo a medical intervention and our side of the bargain should be we want this to be completely safe.

In fact, what we know about vaccines and this is HHS's own studies, a 2010 study by the Agency for Healthcare Research. That was commissioned to look at vaccine injury because CDC for many years had been saying vaccine injury only occurs one in a million. What AAHRQ found, which is a federal agency, they looked at one HMO, which was a Harvard Pilgrim HMO, and a machine cluster analysis. In other words, artificial intelligence, counting. Very, very accurate accounting system and they said the actual rate of vaccine injury is 2.6 percent. That means one at forty people is seriously injured by vaccines.

Do we have a right to say, we are gonna impose this intervention on people, where there's a one in forty chance that you may get injured in order to protect hypothetical people catching that particular disease? For anybody, and this I think is something that Alan really has to, I think Alan, that you need to come to terms with in terms of crafting your own arguments of this, it's not hypothetical that vaccines cause injury and injuries aren't rare. The vaccine courts have paid out $4 billion and the threshold for getting into a vaccine court and getting a judgment, HHS admits that fewer than 1 percent of people who are injured ever even get to court.

The other thing is vaccines are zero liability. This is an industry that went to Congress in 1996 and it had a diphtheria, tetanus, pertussis vaccine at that time that was causing brain injury in one out of every 300 people. And they said to Congress, we cannot make vaccines safely. They are unavoidably unsafe. That is the phrase in the statute unavoidably unsafe. The only reason that we're gonna continue to make vaccines is if you give us complete blanket immunity from liability and Congress gave it to them. So today you have a product that if it injured you no matter how negligent the company was, no matter how sloppy their line protocols, no matter how toxic the ingredients they chose to use, no matter how grievous your injury, you cannot sue that company and that company, therefore, no incentive to make that product safe. And that should be troubling to any of us who are part of the legal system that is saying we are gonna force people to take this intervention.

DERSHOWITZ: Look, I agree with much of what Robert has said. First of all, I completely agree the Supreme Court decision in the Jacobson case in 1905 is not binding on the issue of whether or not you can compel somebody to take the vaccine. The logic of the opinion, however, not the holding, the logic of the opinion and subsequent opinions . . . strongly suggest that the courts today would allow some form of compulsion if the conditions that we talked about were met. Safe, effective, exemptions in appropriate cases. You talked about healthy people being compelled to take a vaccine, which is not designed to help them. Of course, it's also designed to help them. But the major function is to make sure that they don't become typhoid Marys and spread the disease to other people. But when you take a vaccine, you also increase the chances that you will not get the terrible, terrible disease. I think you're gonna have to concede, Robert, that the smallpox vaccine had an enormous positive impact on wiping smallpox from the face of the earth. Smallpox was a dreaded, dreaded, dreaded disease. The Black Plague back in many, many centuries ago, if there had been a vaccine back then could have saved probably millions of lives. We don't know what COVID-19 vaccine will look like, but on the assumption, and here we have a real argument, on the assumption that it would be effective and would stop the pandemic and would cause some injury to some people, then you have to ask how the courts would strike the balance.

When you're a young 18-year-old healthy person, and we have a draft as we had in the Second World War, we don't have it now, but at that point in time, a young 18-year-old was told, look, Congress has given the Army complete exemption. We're not liable if you're shot by the Nazis or by the Japanese, you have to risk your life in order to protect other innocent people in the country. And it's not a perfect analogy, obviously, but it does show that the courts have given to the government, the authority to sometimes make decisions that require you to sacrifice your life.

I have to tell you, I don't wanna become personal about this, but I don't think there's any family in the history of America that has ever made more sacrifices in the public interest than the Kennedy family. It broke all of our hearts to see how much sacrifice the Kennedy family personally made in order to, particularly Robert Kennedy. He put himself in harm's way so many times on behalf of the Civil Rights Movement. People forget how much he put himself in harm's way on behalf of Israel. He was a great friend of Israel, a great supporter of Israel, and the horrible man who killed him, killed him because he was a Palestinian who hated Bobby Kennedy senior's views. Sacrifice is part of the American tradition, and the Americans owe the Kennedy family an enormous debt of gratitude for their sacrifice. Now those were voluntary sacrifices. President Kennedy went to Dallas, knowing there were risks. Robert Kennedy went to Los Angeles, knowing there were risks. By the way, I was working on his campaign. The night I was woken up in the middle of the night to learn the horrible, horrible, horrible, tragic news. And those were voluntary acts. And obviously we're talking about a very different thing. We're talking about involuntary acts, but being drafted as an involuntary act.

Again, to mention the Kennedy family, the oldest brother of the Kennedy family, volunteered to serve in the Army and was killed in combat as a great hero. But there were others who didn't volunteer. Many of my own relatives served abroad. So, we demand sacrifices, and we don't demand perfection. I think both Robert and I agree that we live in an age and it's a terrible time that we live in where everything has become politicized. You mentioned that when we were kids, I remember not being able to swim in the summers of 1953, four or five because of polio. My friend died after being on a lung machine and the blessing that we all made to Salk and Sabin for developing the vaccine. But there were consequences. People took the vaccine and did suffer. In the end though, polio was wiped out. We live in a very divisive age.

Let me mention one other point that I think we should be discussing. Today, the *New York Times* has a very interesting story about who the vaccine will be offered to. The *Times* story is not about mandatory. It's about people wanting it and Robert and I completely agree that the program should begin by giving it only to volunteers. We should only get to this terrible tragic choice issue in the end, if it's absolutely essential that people who don't want to be vaccinated have to be vaccinated to get the kind of herd immunity. We all agree with that, but we live in such a divided time that everything has become politicized. On July 4th, the Reverend Farrakhan made a speech to almost a million people in which he urged Black people not to take the vaccine because we know the history of how Black people were experimented on during the terrible Tuskegee time. And yet Black people and Latino people and people of color are highly vulnerable to the illness. Is that a smart thing for Farrakhan to have urged his community? The number of people of color who have refused or who have indicated a refusal to take the vaccine is, I think, slightly higher, according to the report, than the number of people not of color, who are refusing to take the vaccine. I understand that. I understand the suspicion that our country has generated among people. People don't trust people anymore. I wrote an article in early March, right in the beginning, right at the beginning of this, I wrote an article and the title was "Trust Science, but be Skeptical of Scientists." And at that point, I pointed to two things that were being argued by scientists, including the World Health Organization, which I generally support, saying don't wear masks, number one. And number two, the COVID-19 is not contagious by air. It has no aerosol contagion.

I wrote an article saying, don't believe that. Masks work, number one. If they didn't work, why would so many doctors be using them? And why would it be so necessary for doctors to have them? And second, I don't believe that there's no aerosol transmission. The disease could not have developed so quickly around the world just by touching surfaces. So, I challenged the medical establishment on that. And I turned out, of course, as we all know, to be right, we know there's aerosol transmission. We know masks have an impact, whether they help you who are wearing it, or whether they only help you in transmitting it. But I would like to throw a question out to Robert. I think I know the answer. Robert, would you be against a law that mandated the wearing of masks in public for everybody, even by people who don't approve of the wearing of masks? 'Cause masks don't kill you. They don't pose the risk that vaccine do, but they do deprive you of freedom. Do you think the state, the government, has the legitimate constitutional power to mandate the wearing of masks by people who refuse to wear masks?

KENNEDY: Let me come back to that. Let me address some of the other things. 'Cause I think that's actually a complex question, and I think the science is very controversial on that. Let me address the earlier thing that you said first. One is, this is a rather esoterical discussion and one that I'm not gonna really drag you into other than to say this. The proposition and the theology that smallpox and polio were abolished due to vaccination is controversial. That is not a proposition that is universally accepted. And if you notice all the infectious diseases, whether it was scurvy or tuberculosis for which there were no vaccines, along with puerperal fever and diphtheria and pertussis and measles all disappeared at the same time without vaccination. Now, the CDC actually examined that because it became such a part of the orthodoxy of vaccines that the idea that smallpox and polio were abolished because of vaccines and these other diseases. Johns Hopkins and the CDC, in 2000, did a comprehensive study of that proposition. The study was published in Pediatrics, which is the Journal for the American Association of Pediatrics, which is a readout fortification for vaccine orthodoxy. So, it's a publication very, very friendly and supportive of vaccination. For people who want to look up this study, the lead author is Guier, G-U-I-E-R. And the conclusion of that study is at the abolishment of mortality from infectious diseases that took place during the first half of the 20th Century had virtually nothing to do with vaccines. It had everything to do with sanitation, with nutrition, with hygiene, with electric refrigerators, with reduction in population densities and essentially engineering solution, clean water and good food. And actually that was a guy called Edward Kass, who was the head of Harvard Medical School at that time, who gave a very, very famous speech in which he warned that people who were promoting vaccines and other technologies would try to take credit for those reductions in mortalities from infectious disease, and he said beware of them because they'll try to monetize them and use that to increase their power and their prestige. So, it's something that you might look at it. Again, it's called Guier, G-U-I-E-R.

I agree with you. There was tremendous faith in vaccination during that period. When you grew up, I grew up Alan, we had to be vaccinated and all of them were deemed as necessary. They were feared diseases. Today's kids have to take 72 vaccines. 72 doses of 16 vaccines in order to stay in school. And that explosion of new vaccination came in 1989. Right after the passage of VCA, vaccine act. The vaccine act gave blanket immunity from liability to vaccine companies. And so those companies all of a sudden looked around and said holy cow, now we've got a product where we are completely excused from the highest cost that afflicts every other medical product, which is downstream liability for injuries. That's the biggest cost for every medicine.

Not only that, the vaccines have another exemption that most people don't know about. They are the only medical product that does not have to be safety tested against a placebo. That exemption is an artifact of CDC's legacy as the public health service, which was a quasi-military agency, which is why people at CDC have military ranks, like

Surgeon General and they wear uniforms. The vaccine program was conceived as a national security defense against biological attacks on our country, and they wanted to make sure that if the Russians attacked us with a biological agent, anthrax or something like that, that we could quickly formulate a vaccine and deploy it to 200 million Americans civilians without regulatory impediments. They said if we call it a medicine, we're gonna have to test it and that takes five years to do double blind placebo testing. Let's call it something else. We'll call it a biologic and we'll exempt biologics from safety testing. So, not a single one of the vaccines, the 72 vaccines now administered to our children, have ever been tested against a placebo. And I, in fact, sued HHS in 2016 and said, show me any placebo studies that you have for any vaccines. And they were unable to do so. None of them have been tested and you don't have to sue them like I did.

Anybody can go on their cellphone and look up manufacturers insert. Hepatitis B vaccine, Gardasil vaccine, polio vaccine. You know how many days the current polio vaccine. You know how many days it was safety testing for Alan? 48 hours. The hepatitis B vaccine, the Glaxo version, was four days, The Merck version five days. That means that if a baby they gave that to had a seizure on six, it never happened. If the baby died on day six, it never happened. If the baby got food allergies and were diagnosed two years later, it never happened. If the baby got autism, which is not diagnosed until four years of age, 4.2 years of age, it never happened. Autoimmune diseases, you cannot see those if you have short-term studies and you can't see any risk, if you don't test it against a placebo. My question is, because of that, nobody knows the risk profile for any vaccine that is currently on the schedule. And that means nobody can say with any scientific certainty that that vaccine is averting more injuries and deaths than it's causing. And my question is how in the heck can we be mandating to children that they take a medical product for which we do not know risk? And to me, that is criminal.

And you know, we started this discussion by talking about how do you avoid a whole discussion about mandating vaccines? The way that you do that is you have a transparent process where people see a vaccine is gonna be tested. They see that it's tested fairly against a placebo. That there's long-term tests that are gonna be able to spot all of these difficulties and at its transparent and open, and yet what we've seen from the current group of COVID vaccines is none of that's happening. They're skipping key parts of the test. Moderna vaccine, which is the lead candidate, skipped the animal testing altogether. When they came to human testing, they tested it on 45 people. They had a high-dose group of 15 people, a medium-dose of 15 people and a low-dose group of 15 people. In the low-dose group, one of the people got so sick from the vaccine they had to be hospitalized. That's 6 percent. In the high-dose group, three people got so sick they had to be hospitalized. That's 20 percent. They're going ahead and making two billion doses of that vaccine.

And by the way, people that they test them on, Alan, are not typical Americans. They use what they call exclusionary criteria. They are only giving these vaccines, in these tests that they're doing, to the healthiest people. If you look at their exclusionary criteria, you cannot be pregnant. You cannot be overweight. You must have never smoked a cigarette. You must have never vaped. You must have no respiratory problems in your family. You can't suffer asthma. You can't have diabetes. You can't have rheumatoid arthritis or any autoimmune disease. There has to be no history of seizure in your family. These are the people they're testing the vaccine on, but that's not who they're gonna give them to. What happens? These people are like the Avengers. They're like Superman. You can shoot them with a bullet and they won't go down. What happens when they give them to the typical American? Sally Six-Pack and Joe Bag of Donuts, who's 50 pounds overweight and has diabetes. What is gonna happen then? You're not gonna see 20 percent, you're gonna see a lot of people dropping dead. These people lost consciousness. They had to go to hospital. They had huge fevers. They're the healthiest people in the world. Unlike any other medicine, Alan, that had that kind of profile in its original phase one study would be DOA.

The problem is Anthony Fauci put $500 million of our dollars into that vaccine. He owns half the patent. He has five guys working for him who are entitled to collect royalties from that. So, you have a corrupt system and now they've got a vaccine and it's too big to fail. And instead of saying, hey, this was a terrible, terrible mistake. They're saying, we're gonna order two billion doses of this and you've gotta understand, Alan, with these COVID vaccines, these companies are playing with house money. They're not spending anything on it and they have no liability. So, if they killed 20 people or 200 people, 2,000 people in their clinical trials, big deal. They have zero liability, and guess what? They've wasting our money because we're giving them the money to play with. People like me and people in our community are looking at this process and we're saying whatever comes out of that process, we don't wanna take it 'cause we're seeing how the sausage is made and it's really sickening. No medical product in the world would be able to go forward with the profile that Moderna has.

DERSHOWITZ: Let me just respond 'cause I think we're coming to some common ground here. I have no doubt that transparency and testing is essential. I don't understand why there isn't a placebo testing and other testing later after the initial vaccine. So, there are many phases in a vaccine. We have an emergency now, and we may have to, in fact, develop a vaccine and make it available to people without placebo testing, without diversity testing, we may have to do that, but there's no reason why over time, we can't do the traditional testing, say with polio or smallpox that are now part of our history and have now existed for so many years. Obviously at this point, there's no reason not to be able to do the placebo and the other kinds of human testing.

The article in *The Times* that I referred to made a very interesting point. It said that the people who were most vulnerable to the disease are the people who probably won't be part of the original testing. The testing is, as you said, done, mostly on people who are quite healthy, but isn't there a natural test that occurs?

You say the pharmaceutical industry has nothing to lose, but look at what happened to the pharmaceutical companies that put forward some of the opiates. They have been driven out of business. Their names have been taken off buildings. They are regarded as pariahs in the world today. Certainly, anybody who runs a pharmaceutical company cares deeply about not killing people.

And even if the government doesn't mandate this kind of testing, and even if they give them exemption from financial liability, surely good people, and I think we assume that people who run companies today, I have a friend who's trying to develop one of the vaccines and he's doing it without profit. He feels so strongly about the need to vaccinate people around the world. So, I think you overstated when you say that the people who are developing those vaccines have no concern whatsoever whether people live or die. I think they do have a concern. I think the government has eliminated their financial liability, but would you be sad?

And the other thing is you say there's no testing, but I'm not the expert. I'm not the medical journal reader, but I've read enough medical journals to know that there is a lot of natural testing. You cite some of it, you cite some of the arguments that say that over years people get autism, people get this, people get that. Those results don't come from the initial testing that allowed the product to go forward. They come from great universities, medical schools and public health institutions that continue to test products over time and report to the public the results of those products. Robert.

KENNEDY: Here's the answer. You raised a bunch of questions. One is the opiate people got busted, Alan, and by the way, no, they were not moral people, they knew what they were doing. Their killing of 56,000 American young kids a year, knowing what they were doing. More kids every year that were killed in the Vietnam War. These are not moral companies. And they only got busted because plaintiff's attorneys could sue them.

DERSHOWITZ: I agree.

KENNEDY: And they got the discovery documents and walk them down to the U.S. Attorney's office and said hey, there's criminal behavior here. That can never happen in the vaccine space. You can't sue them. There's no discovery. There's no depositions. There's no class action suit. There's no multi district litigation. There's no interrogatories, nothing. They never get caught out.

Now these four companies make all of our vaccines. All 72 of the vaccines shots that are now mandated for our children. Every one of them is a convicted serial felon. Glaxo, Sanofi, Pfizer, Merck. In the past 10 years, just in the last decade, those companies have paid $35 billion in criminal penalties, damages, fines for lying to doctors, for defrauding science, for falsifying science, for killing hundreds of thousands of Americans knowingly and getting away with it.

Vioxx which was Merck's. Merck's the biggest vaccine producer. Vioxx, which was their flagship product in 2007, was a pill that they marketed as a headache pill that caused heart attacks. They knew it caused heart attacks 'cause they saw the signals in their clinical trials, they didn't tell the American public and they killed between 120,000 and 500,000 Americans who did not need to die. And most of those Americans were people who had rheumatoid arthritis or they had headaches and migraines. They took that pill believing, and by the way, when we sued them, we got spreadsheets from their bean counters where they said we're gonna kill all these people, we're still gonna make a profit, so let's go ahead.

DERSHOWITZ: Nobody can justify that. I agree with you.

KENNEDY: And they ended up. They should have all gone to prison. Instead, they paid a $7 billion fine, but how can anybody, it requires a cognitive dissonance for people who understand the criminal corporate culture of these four companies to believe that they're doing this in every other product that that they have, but they're not doing it with vaccines. They are, and I just wanna answer your other question. No, placebo testing does not take place after the clinical trials. The reason for that is HHS has adopted a very unethical guidance as it is unethical once vaccine is licensed, recommended, it is unethical to do placebo trials or compare of vaccinated versus unvaccinated people. There are scientists who do it, but they're punished for it. It's very difficult for them to publish. They get their funding cut off because nobody wants any study that is going to reveal the truth about vaccine injuries. It just does not happen.

DERSHOWITZ: Look, it's very important that you're making these points because we live in a democracy and nobody is gonna compel a vaccine unless you get democratic approval. Legislatures are gonna have to pass laws doing that. And you should testify about this. Your voice should be heard, but in the end, how do you respond when the American public has listened to you, has listened to your argument, that very persuasive and very convincing and they have an impact on people like me with open minds. And yet in the end, there's a vote by the legislature and the legislature votes to compel vaccinations in the public interest just the way the legislature votes to draft young people to fight wars in which they will die. In a democracy, don't you have to follow the will of the majority?

I agree transparency is all important, and let's shift the debate 'cause you said you wanted me to answer the question. Let's take it out of vaccine for one second because I think it helps analytically. I'm a law professor of 50 years. So, I always do hypotheticals, hypos. So, let's assume the legislature now passes a law. Every 50 States and the United States Congress passes a law requiring everybody to wear a mask when they're outdoors. And you say, well, I'm not so sure that masks are helpful. Maybe they are, maybe they aren't. Congress has hearings. Congress makes a determination that on balance, they are helpful. Wouldn't you agree that it would be constitutional, let's start with constitutional and then desirable, wouldn't you agree that it would be constitutional to mandate the wearing of masks, even if people have political, ideological, medical, religious objections because a, the wearing of a mask is only an inconvenience. Maybe it'll cause a little irritation by some people that will require a topical pharmaceutical, and it has the potential not to save the world, but to improve the possibility of not having communicable diseases. Wouldn't you agree that mandatory mask wearing would be constitutional?

KENNEDY: Well, if I accepted all of your precedence, then perhaps I would. The thing is, I know a lot about the mask. And my organizations, CHD, has not taken a position on them, but I've read at least three med reviews involving hundreds of studies on masks, and the majority of the studies, in fact, there's a BMJ study from 2015, that says that a mask is actually likely to spread the disease and to make you less healthy because the carbon dioxide that you're breathing and the people who wear the mask are more likely to get sick. I'm not saying that that's my position. I'm just saying there's a lot of contrary science out there.

DERSHOWITZ: Do you wear a mask personally when you go out?

KENNEDY: If the science was clear, if the science was clear, then I'd be much more sympathetic to your view. Let me ask you this. Let me just answer the other question you had. You said we have to rely on the majority. Well, I grew up in the state of Virginia, Alan. When I grew up, it was illegal. The majority voted and it was illegal for a Black man to marry a White woman, illegal for Blacks to vote. So, the majority is not. No, in a democracy, you have the courts there that protect our rights.

DERSHOWITZ: I agree, I agree.

KENNEDY: And unfortunately, we are in a situation today where we have tremendous corruption. I don't mean in Congress, which receives more money from pharmaceutical companies than any other industry, pharmaceutical gives in lobby twice the amount that oil and gas, which is the next big one, four times what defense and aerospace. There are more pharma lobbyists in Congress than there are members of Congress in the Senate. So, we have lost, the legislative independence body, Unfortunately, Alan, the agencies are also captured. Now you know about agency capture. It happens everywhere. I have sued EPA my entire life. We just sued the EPA. We just sued Monsanto. We got an historic judgment, a $12 billion settlement in the Monsanto case and I was part of that trial team. One of the things that happened during that trial is that EPA took a position against us. They took the position that glyphosate, Roundup, has caused cancer. As it turns out we got internal memorandum that show that the head of the pesticide division in EPA was actually working secretly for Monsanto and killing studies and twisting studies and ghost-writing studies to falsify the science.

DERSHOWITZ: Look, you're doing great work.

KENNEDY: We were able to show that to the jury. Now imagine this, that's EPA, which is an independent agency. Imagine this, FDA has 50 percent of its budget from vaccine companies, from the industry, 50 percent. The CDC has an $11.5 billion budget. And 4.9 billion of that is buy-ing and selling and distributing vaccines. CDC is a vaccine company. It owns 57 vaccine patents. So, it can make money on every sale of a vaccine. NIH owns hundreds of vaccine patents. NIH owns half the patent for the Moderna vaccine. There's five individuals at NIH, and the rules at NIH, if you're a scientist or an official working on a vaccine you're allowed to collect $150,000 a year in royalties on sales that that vaccine makes. These regulatory agencies are actually vaccine companies. The vaccine marketing sales part of those agencies is the tail that is now wagging the regulatory dog. They are not doing their job as regulators.

And in fact, the senior scientist at CDC today, the senior vaccine safety scientist, who's been, he's still in fact, he was a senior scientist there for 18 years, he is the author or coau-thor on all of the major studies that CDC has produced on vaccine safety and particularly the studies that show the vaccine does not cause autism. His name is Dr. William Thompson. Three years ago, he came forward and he said, we have been ordered to fake all the science of the last decade on autism. And he said, in fact, we were in the major study, which is called DeStefano 2004, the most study on this subject on PubMed. And he said in that study we found out that Black boys, who got the MMR vaccine had a 363 percent greater risk of getting an autism diagnosis in Black boys who waited after three to six months. He said he was ordered to come into a conference room with all that data, with his four other coauthors by their CDC boss, Frank DeStefano, who then ordered them to destroy that data in front of him in CDC headquarters and then published that study saying there is no effect.

So, you have an agency that is really just an arm of industry and the people who are in my community, who are paying who have derided and vilified these mothers who have vaccine-injured children, are being vilified in the press, who are saying, wait a minute, we have read the scientific studies. We have read about the industry corruption. We need to talk about this. They're being silenced by the press. They're not allowed to tell their stories. And nobody is talking, not a single member of Anderson Cooper's staff or Sanjay Gupta has made any effort to talk to Bill Thompson. And he has been begging to be subpoenaed. And he's still at CDC.

DERSHOWITZ: Look, the reason to do this debate is because I think you perform an important function by bringing out some of these ties, some of these connections. You perform an important function when you bring lawsuits against corrupt pharmaceutical companies. But my question is this, knowing all that you know now and putting aside the issue of let's assume we didn't have mandatory vaccinations, let's assume you win that debate. And it's only voluntary vaccinations now. And they come forward with a vaccine that they say will stem the tide of the pandemic. And you're allowed to go on television, on Anderson Cooper. Would you urge all the American people not to take the vaccine? Would you become part of the campaign not to take the vaccine if it were voluntary?

KENNEDY: You know, look. I'm not anti-vaccine. People call me anti-vaccine because of the way I'm marginalizing being silence.

DERSHOWITZ: That's why I am asking you the question.

KENNEDY: I'm not anti-vaccine. Alan, I've been trying to get mercury out of fish for 37 years. Nobody calls me at anti-fish.

DERSHOWITZ: I support you on that, 100 percent, but what would you do? What would you tell the American public if the vaccine were available, and if you were invited say to speak to members of the Black community, members of the Latino community, members of the general American community? And they said.

KENNEDY: If the vaccine, listen. If they come out with a vaccine that it does what it says it's gonna do, you give one shot, you get lifetime immunity and there are vanishingly rare, serious injuries. Well, I don't mind jab site, red nose itching, forget about it. I don't care. I'm talking about deaths, brain damage, one in a million that may be acceptable. In that case, and it works, then I'd say, I'd tell people, yeah, I'm gonna get it. Let's go ahead and get it.

DERSHOWITZ: What if it was one in 1,000, not one in a million? That's more realistic.

KENNEDY: One in 1,000. No, of course, not. I'm not gonna tell one in 1,000 people to die, so that 999 people can avoid COVID, particularly since the hospital rate for COVID I mean a healthy person has basically zero chance dying from COVID. You need to give it to a tremendous number of people to save one life, and the problem with this vaccine is we don't know if the vaccine is gonna kill more people when you start giving it to those people with the comorbidities, 54 percent of Americans now has diabetes, overweight, rheumatoid arthritis. 54 percent of us. I'm not even talking about smokers and vapers. 54 percent of us has chronic disease. They're testing it on one group and they're going to give it to one other and we need to know what the risk factor is in the people that they give it to.

DERSHOWITZ: I agree with that.

KENNEDY: Let me just say, I completely thought it was gonna be made before. I've sued the EPA for many years and it's a captive agency. What would happen if EPA made half of its annual budget selling coal? That's what you got with these regulatory agencies. They're completely corrupt.

DERSHOWITZ: You're performing an important function doing this. Let me ask you another question. What if we had a system which said this, you have two choices. One, you can have the vaccine or two, you can refuse to take the vaccine, but if you refuse to take the vaccine, you have to remain in quarantine until such time as the pandemic is basically passed. So, it's your option. The one option you

don't have. You don't have the third option that is not taking the vaccine and mingling with the public and risking
other people getting COVID, not only young people
although young people do die. The Broadway actor who had
his leg amputated and recently died tragically without any
preexisting conditions. What if we gave people that option?
Quarantine is the option for refusing to accept the vaccine,
but you don't have the third option of refusing to accept the
vaccine and walking around the public without masks.

KENNEDY: That sounds like a reasonable position. The
problem is it's not the way the world works. Let me explain
why. Here's how the world works. And the best analogy is
the flu vaccine. So, a flu vaccine is very much like the coronavirus vaccine. We've had the flu vaccine for 90 years. So,
every year it's fine-tuned and we perfected. And originally,
they told us the flu vaccine you'll get one shot, you'll have
immunity for life. And then it turned out, no, we need to
get it every year.

DERSHOWITZ: Because there are variations of the flu.

KENNEDY: And the same thing is highly likely to happen
with coronavirus. Now the Cochrane collaboration, which
is the ultimate arbiter for vaccine safety, it is the highest
authority and the British Medical Journal have done three
giant meta reviews on the flu vaccine literature. So, they
look at all the literature that exists. The peer reviewed literature that is on PubMed. I think 127 studies. They did it in
2010, 2014 and 2017. Here's what they found. CDC said the

flu vaccine is 35 percent effective. That's what they claim. Cochrane collaboration is that no. You have to give 100 flu shots to prevent one case of flu, number one. Number two, there is zero evidence that the flu shot prevents any hospitalizations or any deaths. Number three, the flu shot transmits the flu. In fact, if you've got a flu shot, you're six times more likely to give somebody else the flu than if you didn't get the flu shot. And this is true, Alan, for many, many other shots for example, the polio vaccine, which you know about is so good at giving polio to other people that 70 percent of the polio cases in the world today come from the vaccine.

DERSHOWITZ: So, let me ask you a specific question.

KENNEDY: And the chicken pox. If you go to the chicken pox manufacturer's insert. It says, if you get this chicken pox vaccine, you should not go near a pregnant woman for six weeks or anybody who is immune compromised. Same with pertussis, you become an asymptomatic carrier. You're not guaranteeing. And, in fact, AstraZeneca vaccine, the Oxford vaccine, which is the other leader, when they gave it to monkeys, the monkeys continued to transmit the disease and Bill Gates and Fauci have been going on TV saying we may get a vaccine that protects you, but you may still be transmitting it. So why are you gonna lock that guy up in a house. People who are now asymptomatic carriers, 'cause they got Gates' vaccine.

DERSHOWITZ: Let me agree with you first of all, if they develop a vaccine that only prevents you from getting it, but doesn't prevent you from transmitting it, I would not be in favor of compelling that vaccine. And I think the Supreme Court would not accept that as a rationale. But I want to ask you a direct question. I'm 81, almost 82 years old. My doctor, who I love and admire, says to me every year come October, you must get the flu vaccine. You must get the vaccine against pneumonia, you must get the vaccine, whatever it is, against shingles. I listened to my doctor, who I love and admire, he has been taking care of me for years. Should I instead listen to you and not take the flu vaccine?

KENNEDY: Nobody should listen to me. People need to do the science themselves. And I would say to you no, listen to your doctor. What Reagan said about Gorbachev. Trust, but verify. You look at the vaccine inserts, Alan. Looked at some of the science and I would say, in a million years I would not take the flu shot. And I'll tell you why, because this is what Cochrane and BMJ have found. People who take the flu shot are protected against that strain of flu. But they're 4.4 times more likely to get a non-flu infection. And you might find, and a lot of people do, that they get the flu shot and then they get sick. They're usually not getting the flu; they're getting something that is indistinguishable from the flu because the flu shot gives you something called pathogenic priming. It injures your immune system so you're more likely to get a non-flu viral upper respiratory infection.

In fact, the Pentagon published a story and you can cite this. It's by Wolfe, W-O-L-F-E. In January of this year, in which they said the flu shot not only primes you for flu, but it primes you for coronavirus. They had a placebo group and they had a vaccine group because they wanted it for military readiness to see if the flu shot was prophylactic against coronavirus. What they found is actually the people who got the flu shot were 36 percent more likely to get coronavirus. And that's not a lone study. We found six other major studies that say the same thing. If you get the flu shot, you're more likely to get coronavirus. This is what the science says and you should not listen to me, nobody should.

DERSHOWITZ: I understand.

KENNEDY: Read the science.

DERSHOWITZ: So, let me understand the implications of your position on the flu shot. Not only would you not take the flu shot and urge me to look at the science and in the end, decide not to take the flu shot because it's too dangerous, but you would also, if I take the implications of your position accurately outlaw flu shot, make it illegal, because in your view and in the view of the scientist you quote, the flu shot causes more harm than good and increases the chances of us all getting the coronavirus. Do I understand the implications of your view correctly?

KENNEDY: Yeah, but I wouldn't take that sort of extreme position. What I would say is we should have vaccines, but we shouldn't have a one-size-fits-all mandates. There may be some situations where even a flu shot would be beneficial to somebody because a flu shot is not completely ineffective. It does probably give you protection against that year's flu strain, if they get it right. And there could be a situation where some-body's life depended on getting that flu shot, but to mandate the flu shot population wide I think is criminal. Look, all you have to do Alan, and this is what Cochran said, is look what's happened to longevity in the elderly, since we started man-dating the flu shot to elderly people. Those are the people; their life expectancy had dramatically gone down as the flu shot proliferated. And if you see the people who died during the COVID crisis, and there's no science on this, but it's observational, it tended to be people who got their flu shots. People who were in nursing homes who all get flu shots. People who are first responders who get flu shots.

DERSHOWITZ: So, with all due respect, I don't under-stand the implications of your position. If you're right, why wouldn't it follow that the flu shot should be illegal? You said it's criminal to mandate the flu shot because it kills people in my age category. So if you had to cast the deciding vote, if you had decided to run for Congress, instead of doing the great work you've done over so many years, and you were the deciding vote in the United States Senate, and there was a bill to outlaw the flu shot, Why wouldn't you vote for it if you think it causes harm?

KENNEDY: I'm kind of a free market guy, I think. You know what, I'm against mandates. I think there may be sit-uations where that product might do some good for some-body, but I just don't believe it should be mandated. I wouldn't think, for example, that Viagra should be man-dated to every human being on the planet, right. But there may be somebody who says I wanna take that medication. Let them do it. I'm not gonna order everybody to do it.

DERSHOWITZ: Look, you and I are on the same page there. I'm curious what you think of this. 'Cause I feel very strongly about this. Let's assume you have a drug or pharmaceutical that hasn't been tested, that is potentially dangerous, but has a 10 percent chance of curing pancreatic cancer in terminally ill patients. Do you agree with me, and with President Trump on this issue, that individuals who are dying should have the opportunity to go off label and to take dangerous drugs that probably will kill them, but increase the chances that they remain alive? That that should be a matter of individual choice.

KENNEDY: I have a big libertarian streak in me. I think people should be left to their own choices wherever possible unless it's gonna do some harm to others. Let me address just one last thing that you were talking about.

DERSHOWITZ: We agree with that, we both agree with John Stuart Mill.

KENNEDY: I think we agree on most of it. You said if it's tested against placebo and this, I think is why people like me, are suspicious, are reticent. The Oxford vaccine, which is the other leader. Gates has a huge investment in it, Fauci is pushing it. It is a leader. AstraZeneca now is branding it. That vaccine is run by a guy called Andrew Pollard, who's at Oxford. A very, very famous, powerful virologist. He originally promised, at the beginning, he said, we're going to test it against the placebo. We're gonna do what's never been done in vaccinology before we're gonna actually use an inert placebo and test it. And then in the middle of his phase two, he said no, we're gonna test it against the meningitis vaccine. The meningitis vaccine is a vaccine with a really high injury profile. It has, listed just on its manufacturing insert are 50 deadly serious injuries, including Kawasaki disease, Guillain-Barre, paralysis, seizure, heart attacks and death and hepatitis and all kinds of autoimmune disease. It's probably, it's arguably the most dangerous vaccine.

So instead of giving his placebo group an inert placebo he's giving him the most dangerous vaccine. Again why? It's a ploy that vaccinologist use. They give their placebo group something that's horrendously dangerous to mask injuries in the vaccine. And so, everybody on my side sees this and they say, he's not being honest. We do not know what the risk profile of that product is. We are never gonna take that product because it was never tested against a placebo. Make them do the science. Don't say to get angry at people who are skeptical and say, oh you're skeptical. We're watching the sausage get made, and it's an ugly process. And by the way, he gave that vaccine to a bunch of monkeys, macaques, and then he challenged the macaques by exposing them to the wild coronavirus.

DERSHOWITZ: Yup. Yup.

KENNEDY: All of the macaques got sick. So, the vaccine doesn't work, but because the British government put 90,000 pounds into it, he now has an order to make two million doses with a vaccine we know doesn't work and they're going forward with it anyway. And he refuses to test it against the placebo. So that gives us zero faith.

DERSHOWITZ: So, let me, first of all, say nobody should be angry at you. People should be praising you for bringing this to the attention of the American public. Let me just summarize if I can, my view, and then you can get the last word. I am thrilled that we had this debate. I think the public watching the debate has learned. We've learned how much we agree about. We're both libertarians. We both agree with John Stuart Mill that the government shouldn't be compelling you to do anything just for your own good, but they can compel you to do things that prevent harm to others. Oh, we have some disagreements about mandates. I think we both agree that any vaccine should start out by being offered voluntarily. We both agree that people should be offered the vaccine initially and take it on a voluntary basis. And that mandatory vaccination, which presents very daunting, moral and constitutional issues should not be required until it's proved absolutely necessary by the consensus of medical opinion. I think we also agree that the First Amendment and the spirit of the First Amendment requires that this debate continue.

And so, I'm pleased that we had this debate. You've persuaded me about some of the medical issues. I will look further into medical issues. I don't think I've persuaded you on the constitutional issues. And I know you haven't persuaded me on the constitutional issues. I still take the position though in a democracy the courts do have the final word that I do believe that if there were legislation mandating in extreme circumstances with safety and other considerations taken into account, mandatory vaccination, I do believe the Supreme Court would and should uphold mandatory vaccination under those circumstances. That's the major area we disagree with. But in practical terms, I suspect we don't have a lot of disagreement that will come to fruition in the next year or so because in the next year the big issue will be how to get the vaccine voluntarily to as many people as possible who are willing to take it. And so, thank you for putting together this debate. I think it really was informative. And thank you Robert, for accepting the idea of debating on this issue.

KENNEDY: Thank you, Alan. I wanna express my gratitude to you on behalf of myself and everybody in this community. People who are called the anti-vax. They're mainly not anti-vaccine. Almost all of them are the mothers and fathers, of intellectually disabled kids who gave all the vaccines, who did what they were told and then their child was injured and that prompted them to go out and do the research. Those people should be allowed to speak. Those people should not be gagged. They should not be considered heretics. They should be allowed to tell their story and they should be treated with compassion and understanding and patience and intellectual openness toward their stories. They shouldn't be vilified. They shouldn't be gaslighted. They shouldn't be ignored and right now, particularly at a point in our history, where we're talking about giving lots of people a vaccine their stories are more important to hear than ever.

I wanna thank you. 'Cause for fifteen years, all of us have been trying to do a debate and we haven't been able to get Peter Hotez to do it. We haven't been able to Paul Offit, Ian Lipkin, any of the leaders have been scared to sit where you are now. And I want to thank you so much on behalf of all of us, but also our democratic traditions for coming here. Thank you, Alan.

DERSHOWITZ: Well, thank you Robert.

Chapter 6
Censoring Debate about COVID and Vaccines

M y debate with Robert Kennedy was seen by thousands of viewers on YouTube. I received numerous emails and calls, some agreeing, others disagreeing. Then YouTube took it down. It can be seen on Valuetainment and has been viewed nearly two million times. When YouTube took it down, I wrote the following op-ed.

Last summer Bobby Kennedy, the distinguished environmental lawyer, and I had a thoughtful and substantive debate about the constitutionality of compelling people to be vaccinated against COVID.

Many people watched the debate on YouTube and commented on its educational value. Both sides were presented fairly and effectively, and viewers were able to decide for themselves who got the better of the argument. But that will no longer be possible, because YouTube has emailed the following to Children's Health Defense: "Our team has reviewed your content and unfortunately we think it violates our medical misinformation policy. We have removed the following content from YouTube: Kennedy and Dershowitz debate."

YouTube did not disclose the reason why they believed the debate violated their medical misinformation policy. Nor did they specify what was "misinformation." Surely the debate over the constitutionality of mandatory vaccination did not contain medical misinformation. In my portion of the debate, I provided no medical information, so I could not have provided "misinformation."

If YouTube believes that Kennedy's part of the debate contained medical misinformation, they should specify precisely what constitutes such misinformation, so that Kennedy can either provide documentation or remove the offending material. Instead, they simply took down the entire debate, thus denying their viewers the opportunity to participate in the open marketplace of ideas regarding the important issue of compelled vaccination.

I oppose all censorship of ideas by YouTube, but I oppose even more strongly the censoring of debates that present all sides of an issue. Such debates are the heart and soul of American democracy. If YouTube had been in existence at the time of the Lincoln-Douglas debates, would they have taken them down because they contained some offensive racial references? Would they have taken down the debates over the ratification of the Constitution because they contained justifications of slavery? Are they going to take down other current debates about the source of the virus, about mask mandates, or the opening of businesses? Where will it stop?

By taking down our debate while leaving up many false statements—medical, historical, and scientific—YouTube conveys the impression that they somehow validate the truth of everything they don't take down. By thus implicitly attesting to the truth of these many falsehoods, YouTube itself would seem to be violating its own policies. That is part of the problem with censorship: Either it censors selectively, thus validating what it doesn't censor, or it censors pervasively, thus keeping much valid information from the public.

Consider for example recent social media postings that compare certificates showing that a person has been vaccinated—so called "vaccination passports"—to the yellow Star of David that Jews were required to wear in areas under Nazi control during the Second World War.

Among the people making this bizarre, bigoted, and ahistorical comparison are Congressional Rep. Madison Cawthorn, a freshman Republican from North Carolina. He has said, "Proposals like these smack of 1940s Nazi Germany." The Libertarian party of Kentucky has asked the following rhetorical question: "Are the vaccine passports going to be yellow shaped like a star and sewn on our clothes?" Ambassador Richard Grenell, former president Trump's ambassador to Germany, who is now a member of the US Holocaust Memorial Council, tweeted a meme that showed a Nazi soldier accusing someone of "hiding unvaccinated people under your floorboards." This was based on a quote from the film *Inglorious Bastards*.

In Great Britain, a right-wing writer named James Delingpole published the following tweet: "Wouldn't it be better to just cut to the chase and give unvaccinated people yellow stars to sew prominently on to their clothes?"

These ignorant and bigoted comparisons are not-so-subtle forms of Holocaust-denial: They imply that the Holocaust was nothing worse than allowing vaccinated people to have a certificate, or even denying unvaccinated people the right to infect others. Compelling Jews to wear the yellow star was designed to identify them for transport to death camps where they, their children, and parents were forced into gas chambers and murdered. A vaccine certificate is a symbol of life not death.

Whether one agrees or disagrees with vaccination certificate or passports, no analogy should ever be made between such public health documents and symbols of death during the Holocaust. Yet, despite my outrage over this analogy, I would not ask YouTube to take down these hateful posts.

In America we debate, we disagree, we argue. We tolerate bigoted rap songs, hate speech, even Holocaust-denial. We don't censor.

Social media are private businesses not governed by the First Amendment. They have the right to do the wrong thing, including to censor. We have the right to object to their doing the wrong thing by speaking out against them. I am exercising that right in this column.

Chapter 7
Does Bodily Autonomy Extend to One's Organs after Death?

The case for bodily autonomy has been extended by some beyond the life of the body. Several years ago, I wrote an answer to this claim that may be relevant to the current vaccination debate about the inherent right to prevent the state from intruding on one's body. Although there is a difference between a live person and their dead body, it is the live person (or his relatives) who decides whether his organs shall be removed and donated after their death. Here, with some changes and updates, is what I wrote.

At this point in our history, human beings have a right, recognized by law to be buried along with their organs even when these organs could be used to save the lives of other human beings. There may come a time in the future when people who are dying of organ failure may claim a right to use the organs of dead people. How would a moral society evaluate this claimed right? How would the constitution deal with it if mandatory organ donation were enacted by a legislature? Before we seek to answer these questions, it will be useful to look more broadly at the role of social institutions in dealing with human nature, particularly as it relates to the human body.

Law, religious custom, tradition, and morality all share in common certain mechanisms for influencing and improving human conduct—for making it less "natural." These mechanisms are premised on the assumption that in the absence of external rules of conduct, most humans would tend to act selfishly (I define selfish to include family). The rules are designed to discourage human beings from making individualized ad hoc decisions based on a selfish cost-benefit analysis of the particular situation confronting them. Instead, they impose on individuals the obligation to think more generally, more broadly, more categorically, more altruistically, and more communally—that is, more morally. These rules prohibit different categories of acts. Some prohibit core evils, such as the killing of innocent people. Others prohibit acts that are not in themselves immoral but that are thought to lead to core evils. Such prohibited acts including driving too fast or while drunk. Yet other rules seem designed simply to condition people to accept limitations—even artificial limitations—on their appetites or instincts. These include ritual restrictions on the eating of certain foods or the performance of certain ritually impure acts.

The rule of law, religion, and morality seek to make it more difficult to act on the instinct of selfish preservation of individual and family and to make it acceptable—indeed, obligatory—to act on the basis of a broader principle. The specific principle may vary, depending on whether one is a Kantian, a Mill utilitarian (act or rule), a believer in the Bible, or a follower of any other set of rules, but the mechanism is similar: It requires you to act not as if yours were the only situation, but as if it were part of a principled set of mutually binding obligations.

Consider, for example, the issue of cannibalism. Start first with the eating of a human being who has already died. Absent the constraints of law, morality, religion, and so on, any rational starving person—say a sailor in a lifeboat, a soldier lost in a jungle, an entire city besieged and surrounded—would not think twice about eating the fresh meat of a dead person, any more than he or she would think about eating the fresh meat of a dead animal. Some might argue that it is "natural" for human beings to be revolted by the thought of eating the flesh of fellow human beings, even if they were not responsible for their death. But throughout history and throughout the world, people have eaten dead humans. We are revolted by the thought because law, morality, and religion have conditioned us to become revolted. If we had grown up in a world in which the eating of human flesh was common we would not be revolted by such a diet any more than by our diet of animal flesh Perhaps someday when artificial food becomes an easy alternative, our great-grandchildren will be as disgusted by the prospect of eating the flesh of animals who were once alive as my grandparents and parents were revolted by the thought of eating certain dead animals such as pigs and lobsters.

Why, then, do we not eat human flesh? For some, the answer is simple: God has told us not to. But the gods of some Polynesians said it was permissible. What if our God had said it was permissible? Putting the same question at a different level of abstraction. Why did our God—or those who have purported to speak in his name—single out the flesh of humans as prohibited food? It seems a waste in a world in which so many are starving. Perhaps the answer lies in the slippery slope. If we would permit the eating of the flesh of someone who was already dead, we might be more inclined to kill them for their food value, just as we do with animals. So, we create a prophylactic rule—or to use the words of the Talmud, we build a fence around the core prohibition. The core prohibition is the killing of human beings; the fence is the prohibition against the eating of already-dead human beings.

Perhaps there is another core principle behind not allowing the eating of human flesh. Is it that somehow the human body is sacred? That it should never be used as a means toward the end of saving another human life? Surely the answer to these questions must be no, as evidenced by the fact that we do not prohibit the harvesting of body organs of dead human beings for transplantation into live human beings who might otherwise die for lack of a needed organ. In principle, what is the difference between "harvesting" the flesh of dead human beings to save the lives of other human beings and "harvesting" their other organs? It cannot be personal preference alone. If it were, I might personally reject the distinction unless someone could make a persuasive argument in support. If I were dead, I would just as soon have my flesh eaten in order to save the life of another human being as have my heart or kidneys removed for transplant. I make no claim to ownership of my body once dead, as evidenced by the fact that I have signed on as an organ doner. I would also like my body to be autopsied if that would provide useful disease-preventing information to my children and grandchildren. If there were a place to sign on as a flesh doner, I would do that as well—unless a larger principle were at stake.

There is, of course, this difference between transplanting an organ and eating the flesh. The organ is generally needed to save life. There is a one-to-one correlation. Eating human flesh on the other hand, could become an appetite rather than a necessity. Indeed, we accept eating of human flesh when absolutely necessary to save life as in shipwrecks and following the famous airplane crash in the Andes in 1972. We just don't want it to become routine. We might develop the same attitude toward organ transplants if people began to transplant the blue eyes of dead people for purely cosmetic reasons.

Even—perhaps especially—when organs are needed to save lives, we properly worry that transplantation may encourage the killing of some human beings for their organs. Such practices are believed to exist in certain parts of the world even today, and we have built fences to protect the living from being killed for their organs. No moral, religious, or law-abiding person would order an organ if he knew someone would be killed to provide it. If we chose, we could build an even higher fence: namely, prohibit the use of the organs of the dead, just as we prohibit the use of their flesh.

When organ transplantation became possible, some religious groups made precisely that argument: the human body is sacred; it must be buried with all its organs; removal of any organ, even if necessary to save human life, is a desecration. That is no longer the position taken by mainstream religions, most of which now tolerate, or encourage, organ donation (some encourage only the receipt of organs not their donation but that is an unacceptable selfish moral position). Moral leaders should encourage their followers to think of their corpses as containing living recyclable parts. This change in perspective should be made in the interests of saving human life, thereby enhancing rather than diminishing its value. A dead body whose usable organs have been removed should become a symbol of respect for the living body. It is all a matter of how we view it and what we teach our children. There is nothing "natural" or "un-natural" about cutting up a dead body to give life to a live one whether by using its heart or its flesh.

To encourage respect for the living, we mandate respect for the dead. It is not so much that a dead person has rights in his remains, as that the living have rights to see the bodies of their loved ones treated with respect. It is a crime to desecrate a cemetery or a corpse. We require our pathologists to perform autopsies in a dignified manner. We dispose of body parts with respect. Soldiers risk their lives to recover the bodies of their fallen comrades. We do all this not because it matters to the dead, but because it matters to the living. We have learned the lessons of history that teach that societies that disrespect the dead bodies and resting places of the deceased tend to devalue the living bodies—the lives—of their contemporaries. What constitutes respect—burying a body with its organs or without them—is a matter of education and nurture rather than divine law or nature. In some societies, respect for the dead requires that the body be taken to a remote hilltop so that its flesh may be consumed by birds of prey. The circle of life!

The same can be said about abortion. Some who argue against abortion say that if we trivialize the "death" of a living human fetus, it becomes easier to devalue the life of a baby, a mentally retarded person, a prisoner, a Jew, a Black person, an enemy, a stranger. Others argue that to compel a woman to bring an unwanted baby into the world devalues the life of the child and the welfare of the mothers. Again, there is no one naturally correct answer for all moral people.

Another, less compelling example of a fence around the core violation would be in the prohibition against selling and trading ivory. There's nothing wrong, in principle, with using the tusks of dead elephants. But once a trade in ivory becomes acceptable, live elephants will be killed for their tusks. Accordingly, we try to make ivory an immoral and illegal commodity. Likewise, with those who would try to make the wearing of animal fur unacceptable. Again, we can distinguish in principle between stripping the fur from dead animals and killing animals for their fur, but the lesson of history is that permitting the former will encourage the latter. Thus, we see the same principles in operation once again: We impose a seemingly irrational prohibition against a harmless use of resources—the flesh of dead people, the tusks of dead elephants, the fur of dead animals—in order to discourage a violation of the core principle namely, killing to secure these same commodities.

There are, of course, intermediate approaches. We could impose harsh punishments on those who kill elephants for

their tusks while encouraging the use of tusks from elephants that died naturally. Experience would then show
whether it is necessary to have a blanket prohibition on the
sale (or use) of all tusks in order to prevent the killing of
elephants. Or we could distinguish between the use of fur
from animals specifically bred and raised for their fur and
from animals in the wild. In the end, it will be experience
rather than some abstract natural rule that will determine
how high we need to build the fence in order to protect the
core value.

The very concept of a fence is a recognition that rights
are built on our experience with wrongs. It is this experience
that cautions us about the dangers of the slippery slope—
about the inclination of some people to take arguments to
the limits of their logic and beyond. The irony is that it is
experience with moral relativism, situational ethics, and
continua that leads some to argue for absolutes and clear
lines, and to pretend (or persuade themselves to believe) that
these absolutes and lines come from God or nature.

The argument for absolutes and clear lines, rather than
for continua is a plausible one, based on human experience.
It is played out regularly in our courts, as some judges read
provisions of our Constitution as absolute prohibitions on
government power while others read exceptions and a rule
of reason into these same provisions. Justice Hugo Black
argued that the First Amendment's statement that Congress
"shall make no law . . . abridging the freedom of speech"
meant what it said: No law means no exceptions. Justice
Felix Frankfurter argued for a rule of reason pursuant to
which the government could make laws abridging certain

genres of dangerous or offensive speech. When a government lawyer would argue for an exception Black would take out his worn copy of the Constitution and read "Congress shall make no law . . . ," banging the table as he shouted the word "no." Frankfurter would then mock Black by opening his copy of the Constitution and reading the same words, except that he would bang the table when he shouted the word "congress," emphasizing the fact that the literal prohibition applies only to one branch of the federal government, not to the states or to the executive or judicial branches.

Black was a legal positivist and a pragmatist. He did not believe that the absolutes he insisted on came from God or nature. Instead, he thought the framers had decided to impose absolute prohibitions on certain governmental actions as a result of their negative experiences with judicial discretion and slippery slopes. Frankfurter placed greater trust in elite judges and in their ability to interpret constitutional prohibition in a reasonable manner. Both were products of their own very different experiences. Black as a populist legislator, Frankfurter as an elitist professor.

The debate over whether absolute prohibitions or relative continua provide better protections against slippery slopes should be an empirical one that can be resolved by human experience rather than by the Delphic voices of God or nature.

Let us now try to apply the experimental approach to a specific set of questions relating to organ donation. A friend of mine recently died because he was unable to get a suitable heart for transplant. No healthy hearts were available at the time he needed his transplant, and so in order to remain alive, he had to settle for a heart of a patient with hepatitis. The heart transplants worked, but my friend soon died of liver failure.

My friend, unfortunately, is among the large number of Americans who needlessly die each year because other Americans selfishly refuse to donate lifesaving organs after their own deaths. In the United States, there is a presumption against organ donation at death, which can be overcome only if the potential donor has made an affirmative decision to consent to having his or her organs removed upon death. In many European countries, the presumption goes the other way: All people are presumed to consent to the lifesaving use of their organs unless they explicitly take action to withhold consent. The result is that many more organs are available for transplant patients in European countries than in our own.

I can imagine few more selfish and immoral acts than insisting that your lifesaving organs must be buried with you so that worms can eat them, rather than allowing them to be used by other human beings to save their lives or to restore sight. Yet many Americans refuse to consent to organ donation upon death. A significant number justify their act of selfishness by reference to their religion. But what kind of religion would preach that it is wrong to help save lives by donating organs from a dead body? Religious leaders should be in the forefront, urging their followers to overcome their fears and superstitions and take the simple step that will directly save lives.

But religious leadership alone will not eliminate the critical shortage of organs. We need to change the law. At the very least, we should move toward the European system of presuming consent in the absence of explicit withholding of consent. Even this shift of presumption may not produce enough organs. The time has come to raise the question of who owns a person's life-giving organs after that person has died. Do you have a right to have buried or cremated parts of your body that could keep other people alive? Would it violate the rights of dead people or their families for a state to pass a statute mandating organ removal and reuse after death? (There already are statutes requiring the removal and preservation of organs when autopsy is mandated for evidentiary purposes.) Would there have to be an exception for religious objection? These are questions we ought to begin debating. Improvements in medical technology require us to rethink old attitudes about our bodies after death. Treating the dead body with respect is an important element of humanity, but the forms of respect may vary. We as a society might well—and should—come to believe that retrieving organs that can then be kept alive and given to others is a proper way of showing respect.

When organ transplants first became feasible, many traditionalists objected—on moral and religious grounds—to playing God and tinkering with nature. Over time, attitudes changed, and almost nobody today turns down a lifesaving organ on religious or moral grounds. The Golden Rule—which is central to Judaism, Christianity, Islam, and other religions—requires that we treat our neighbors as ourselves. Anyone willing to accept a transplant must be willing to give their own organs. Religions that permit their adherents to receive transplants must permit them to donate organs lest they be accused of hypocritically violating the Golden rule. Perhaps an additional encouragement to transplant donation would be a rule excluding all adults who had not consented to donating their organs from receiving the organs of others. At least, there should a be a preference for those who were willing to donate organs.

Anyone who refuses to sign the box on the driver's license application which constitutes consent to removal of organs after death is either a coward, a fool, a knave, or a slave to superstition or religious fundamentalism. There is no softer way of putting this. It is simply wrong to waste the organs of the dead when they can be used to save life. It is understandable that some relatives of a crash or shooting victim would not be willing to consent to the removal of organs from the bodies of their recently deceased loved ones. But it is not understandable for an adult to refuse to consent in advance to the lifesaving use of his own otherwise useless organs. We should make such selfishness unacceptable as a matter of morality and perhaps even as a matter of law.

Eventually our experiences with organ transplantation may move our society toward the recognition that there should be no right to refuse to have your organs harvested for lifesaving use after your death. There should, of course, be a right not to have your death accelerated in order to maximize the chances that your organs can most effectively be used. That should be a matter of choice. And there should, of course, be a right not to be killed in order to have your organs used to save the life of a more important or wealthier person. We might need fences around these core principles, but there is the danger that a fence built too high may endanger other core values.

Organ transplantation provides a good example of important values clashing with others. First, there is the claimed value in preserving intact the bodily integrity of a dead loved one, or even the right of the dead person to dispose of his body as he chooses. On the other side of the ledger is the value of preserving the life of the person in need of an organ. Does a live person have the right to the organ of a dead stranger if that organ and that organ alone means the difference between life and death? What if the person in need of the organ is a scientist on the verge of curing cancer, the president of the United States, or a single mother of two children? Does their right to live—and our right to have them live—outweigh the right of the deceased to be buried with all his organs? Will it really make a difference if he is buried with one less kidney, no liver, or heart?

Of course, much depends on how the issue is framed. Instead of describing the choice as between the life of the recipient and the wish of a dead person, it could be described as between the right of a human being to make important decisions about the disposition of his body and the power of the state to compel that person to violate his religious, moral, or aesthetic principles. Thus, the framing of rights issues exerts a powerful influence on the moral and political debate and can be used to tilt the debate toward a favored position. Some religious and political leaders are particularly adept at this and advocate on all sides of contentious issues employ these framing tactics.

In general, our society gives individuals considerable authority to direct the disposition of their property after their death, but this right is not without some limits. A husband may not completely disinherit his wife. Nor may he deny the government its statutory share of his property— namely the inheritance tax. A body is, of course, different from a bank account or even a valuable painting. Even with regard to a painting, however, there may be some limits. If a private person who owned an important collection of early Picasso paintings maliciously decided to have them destroyed upon his death, some societies—France, for example— would prohibit such a destruction of what is deemed a national treasure. In the United States, some privately owned buildings are declared historical landmarks and may not be destroyed, even if destruction is in the private financial interests of the owner. All governments assert the power of eminent domain over property needed for certain public purposes.

There may come a time when our collective experience with organ transplantation causes us to disregard (or give less weight to) the wishes of dead persons and their families if their organs could save lives. If experience shows that widespread organ retrieval saves numerous lives at little cost—psychic, moral, and financial—a consensus may emerge regarding lifesaving organs as "rightfully" belonging to those most in need of them. If that were to occur, many more people would probably begin to donate their organs, and there would be less need for mandatory confiscation. But if there were still a shortage, a mandatory system might be considered.

Experience could, on the other hand, move us in the opposite direction. It could turn out that more frequent organ transplantation produces negative consequences such as hastening the deaths of such people with needed organs, diminishing the value we place on the human body, discouraging research on other lifesaving techniques, or creating a caste system in which certain recipients are preferred over others. If this were to be our experience, then it might well reinforce current attitudes against having one's body "cut up" after death. The point is that there is no natural or divinely mandated way of showing respect for the human body. Deciding whether to regard the dead human body as an integral entity to be buried intact, rather than as an expired container for recyclable lifesaving organs, is very much a matter of experience. As experiences change, so do attitudes and so do rights—even rights as emotionally laden as the claimed right to be buried or cremated with one's organs intact.

Chapter 8
Is There Always a Right Answer?

In the real world, there isn't always one right answer to complex moral, scientific, and legal issues. Several years ago, I wrote about that conundrum, which may be relevant to the current debates regarding COVID.

What happens when rights clash? For those who believe in the inerrancy and absoluteness of natural law, both sides of a moral—or right-based—argument cannot be correct. But in real life there may indeed be both a right to life and a right to choose, just as there may be a right to life and a right of self-defense. Why is it supposed that rights may not be in intractable conflict? This conflict may require an agreed-upon process for reaching a workable resolution to the clash of rights within a pluralistic democracy.

An old story about a wise Eastern European rabbinical judge illustrates this reality. The rabbi was hearing a dispute between an estranged husband and wife. The wife argued that her husband had violated her marital rights by sleeping with other women and refusing to give her enough money for necessities. The rabbi listened and ruled, "You are right, my daughter." Then he heard her husband's claim that his marital rights had been violated by his wife's refusal to sleep with him and to cook his meals. The rabbi listened and ruled, "You are right, my son." The rabbi's student interjected, "But rabbi, they can't both be right"—to which the rabbi responded, "You too are right." The husband and wife can both be right—and wrong. Life is more complex and nuanced than language. There may be multiple "rights" in any situation, and they will sometimes conflict.

In *The Brothers Karamazov*, Fyodor Dostoyevsky posed a wonderful moral dilemma to which there is plainly no single right answer. Mikhail, a highly regarded town official who is married and has children, confides in Zozima that fourteen years earlier he killed a woman in a well-planned act motivated by jealousy. The peasant who was falsely suspected of the crime died before he could be brought to trial. No one would materially benefit from Mikhail's belated confession, but his innocent wife—whom he married after the killing—and his young children would suffer grievously: "My wife may die of grief"; "My children, even if they are not stripped of rank and property, will become a convict's children, and that forever"; "And what a memory I shall leave in their hearts." Mikhail and Zozima cite conflicting scriptural and philosophical sources in support of confessing or remaining silent.

If he could join this debate, Kant would surely argue that truth is all that matters, regardless of the consequences. Jeremy Bentham would point to the accumulated unhappiness that would result from a selfish confession—selfish because, according to Dostoyevsky, it would be designed to bring salvation to Mikhail's conscience at a high cost to his family. Other philosophers would invoke principles pointing in different directions, some arguing that consequential thinking would produce great evils in other cases (e.g., killing a "worthless" witness in order to spare innocent family members the grief of their breadwinner's being caught) with others arguing that suffering in silence or even suicide would be a more noble and moral act for Mikhail than shaming his family to assure himself paradise. The question is not which of these solutions—or others—is preferable. The question is whether there is only one right solution, and if so, what is its authoritative source and how shall we deal with powerful and persuasive counterarguments? I believe that reasonable people—moral, religious, and good—can and should disagree about this and other complex moral conundrums, and that we should not presume there is a single, perfect answer to be discovered if only we could access the proper source.

Another example, inspired by Dostoyevsky's Grand Inquisitor scene, also tests the principle that truth is absolute. Imagine being on a visit to the Qumran caves outside Jerusalem and coming upon a previously undiscovered Dead Sea scroll, much older than the ones from the time of Jesus, that contains the account of a meeting of religious leaders deciding how to deal with rampant disbelief, lawlessness, and violence. They come up with the idea to stage a "revelation" on a mountain called Sinai from which God will "give" ten commandments carved on two tablets. They argue about the content of the commandments and finally compromise on the ones we know. The scroll contains considerable details about the staging necessary to persuade the common folk that the revelation is genuine. A later scroll, also newly found, suggests the staging of a crucifixion and resurrection of a man who they will claim was God's son. Someone suggests that he be born to a virgin, as is common in Greek mythology, and that he be crucified and then miraculously resurrected. Calvary is selected as the site for this final miracle. Other scrolls could contain similar staged miracles involving Muhammad and Joseph Smith.

I mean no disrespect to religion by these made-up scenarios; they are merely designed to test principles. You, the finder of the scroll, are a deeply religious person, equally committed to truth. You also believe that organized religions, particularly the Judeo-Christian fathers, are extremely valuable and central to the lives of millions, including your own parents and grandparents, whose faith would be shaken by the disclosure of your findings. You are convinced beyond any doubt that the scrolls are authentic and that the events reported by the Gospels or other holy books were completely fabricated by people of utter goodwill—they were truly "pious frauds." Do you disclose your findings? Remain silent? Destroy the scrolls?

A more mundane variation on this unrealistic scenario occurs whenever a religious leader—on the eve of delivering a major sermon on the importance of believing in God—suddenly begins to doubt God's existence. This surely happened to many rabbis, priests, and ministers during the Holocaust. Does truth require cancellation of the sermon, revelation of the new belief, change in tone, or simply a crossing of one's fingers as the original sermon is delivered?

These are daunting questions—and there are many others—to which a single correct answer would be insulting to the complexity and diversity of the human mind and experience. It is the brilliance of great literature that it is never satisfied with a single, correct answer to a complex and often ambiguous human dilemma. Its characters—through their internal dialogues as well as their external conflicts—reflect the diversity of human experiences and emotions. A great moral philosopher must have the insights of a poet.

Even schematic hypothetical situations—carefully crafted for their one-dimensional simplicity—defy singular answers. The famous "trolley track" dilemmas are designed to show there is no right answer: You are a trolley conductor whose brakes have failed. You see a fork in the tracks ahead. If you turn right, your trolley will hit a group of children; if you turn left, you will hit a single drunk; if you fail to choose, random forces will choose for you. A complicating variation has a straight track leading directly to the children, but there is a turn you could take that would lead you to the drunk. Under this variation, if you do nothing, the children die. You must actively choose to kill the drunk. Countless variations on these "tragic choice" dilemmas are imaginable. Each has multiple "right" and "wrong" answers.

The difficult issue—of morality, legality, and practicality—is how to devise an acceptable process for resolving such conflicts in a pluralistic democracy committed to balancing the preferences of the majority against the right of minorities. There can be no absolutely perfect resolution to these conflicts. They reflect deeply felt moral concerns, intuitions, historical experiences, and worldviews that may and do differ over times and cultures. Recall that as recently as two centuries ago—a blip on the timeline of recorded history—most thoughtful and decent people honestly believed in the moral inequality of Whites and Blacks, men and women, Christians and "heathens," heterosexuals and homosexuals, as well as other dualities. Who can know which of our contemporary moral beliefs—for example, the distinction between the value of human and animal life— will seem unacceptable to our progeny in generations to come?

Law, morality, and even truth are ongoing processes for resolving conflicts in a democracy comprising people with different histories, experiences, perceptions, value hierarchies, and worldviews. To expect one "correct" or one "true" moral answer to emerge from such different backgrounds is to devalue our diversity. We respect our heterogeneity when we construct democratic processes for compromise and for accommodating the living with the inevitable differences— even about rights—that are inherent in such a diverse society.

It should not be surprising that the United States has served as the great laboratory for constructing these conflict–accommodating processes, since our population is the most diverse in the history of the world. We began as a community of immigrants and dissenters with a relatively narrow range of ethnic backgrounds. By the end of the nineteenth century, these backgrounds had diversified, along with our religious differences. Unlike some other countries, which share a more unified tradition with regard to the substance of morality, law, and rights, we have developed a consensus about the processes for resolving our substantive differences. Our system of checks and balances may often result in deadlock or only gradual change, but it is a system designed to accommodate differences. Compromise has been the essence of the American experience, even in areas as difficult to compromise as family conflict and religion.

A recent case decided by the Supreme Court illustrates how we tend to resolve intractable moral conflicts by resorting to process and procedure. The case pitted the rights of the parents of two young children against the rights of the children's grandparents. A woman named Tommie and a man named Brad had two children. They never married and eventually separated. Brad moved in with his parents and brought his children to their home for weekend visits. Two years later Brad died, and Tommie married Kelley, who then adopted Tommie's two children. The children's paternal grandparents—the late Brad's father and mother—wanted to keep seeing their grandchildren on a regular basis (two weekends per month and two weeks during the summer), but the parents wanted them to have less time together (one daytime visit per month). The Washington State statute empowered the courts to order whatever visitation rights would serve "the best interests" of the children, without regard to the wishes of the parents. Pursuant to that statute, a lower court judge granted the grandparents' petition for broad visitation rights, citing his own familial experiences: "I look back on some personal experiences . . . We always spen[t] as kids a week with one set of grandparents and another set of grandparents, [and] it happened to work out in our family that [it] turned out to be an enjoyable experience. Maybe that can, in this family, if that is how it works out."

The stage was thus set for petitioning the United States Supreme Court to resolve this conflict of alleged rights:

those of the parents (in this case one biological, one adoptive) against those of the grandparents (both biological, through the dead natural father).

There is not a word, syllable, suggestion, or innuendo in the Constitution that controls, or even informs, this conflict. Nor were there any binding constitutional precedents or history on this point. The high court has, of course, talked about "parental rights" in relation to the state, the schools, and other outside institutions, but not in relation to grandparents who also claim some parental or family rights. Reasonable people can and should disagree—as a matter of policy—on what the correct answer is. The Washington legislature, presumably after considering all sides of this issue, came down in favor of limited grandparental rights as forty-seven other states have also done. That should resolve the question as a matter of constitutional law. Where there is no constitutional prohibition on a particular answer and when a state legislature acts reasonably in arriving at an answer the Supreme Court has no business second-guessing the state's answer. Only an unreconstructed judicial activist—a judge who believes in substituting his or her own personal moral philosophy for that of duly elected legislators— would consider striking down the Washington statute, or the statute of another state that came to the opposite conclusion and prohibited grandparents from visiting grandchildren over the objection of parents. Supreme Court Justice Louis Brandeis, a paragon of judicial restraint understood that states must be accorded considerable flexibility so that they can be "laboratories" of social experimentation. States should be free to come to differing conclusions on divisive and controversial moral and psychological issues not governed by the Constitution, so long as they do so in a reasonable manner. Did the Washington State statute meet this criterion? That is the question addressed by the justices. The Supreme Court began its analysis by looking to experience:

> The demographic changes of the past century make it difficult to speak of an average American family. The composition of families varies greatly from household to household. While many children may have two married parents and grandparents who visit regularly, many other children are raised in single-parent households. In 1996, children living with only one parent accounted for 28 percent of all children under age 18 in the United States. . . . Understandably, in these single parent households, persons outside the nuclear family are called upon with increasing frequency to assist in the everyday tasks of child rearing. In many cases grandparents play an important role. For example, in 1998, approximately 4 million children—or 5.6 percent of all children under age 18—lived in the household of their grandparents.
>
> The nationwide enactment of nonparental visitation statutes is assuredly due, in some part, to the States' recognition of these changing realities of the American family. Because grandparents and other relatives undertake duties of a parental nature in many households, States have sought to ensure the welfare

of the children therein by protecting the relationships those children form with such third parties.

Despite this experiential basis for allowing grandparental visitation to be ordered by the courts, a plurality of the US Supreme Court struck down the Washington statute on the ground that it was too broad and open-ended. The court also cited experience to challenge any ideal conception of child rearing:

> In an ideal world, parents might always seek to cultivate the bonds between grandparents and their grandchildren. Needless to say, however, our world is far from perfect, and in it, the decision whether such an intergenerational relationship would be beneficial in any specific case is for the parent to make in the first instance. And, if a fit parent's decision of the kind at issue here becomes subject to judicial review, the court must accord at least some special weight to the parent's own determination.

Because the Washington statute did not accord sufficient deference to the wishes of the custodial parents and because it authorized "any person"—not only grandparents—to petition for visitation it was held unconstitutional. The high court thus evaded the direct clash of morality between claims of parents and grandparents and decided the case on technical legal issues, leaving open the question of whether a narrowly tailored law granting some visitation rights to grandparents, over the objection of parents, would be upheld.

The most interesting opinion in the case was written by Justice Scalia, who personally believes that only parents have a God-given natural right to raise their children.

> In my view, a right of parents to direct the upbringing of their children is among the "unalienable Rights" with which the Declaration of Independence proclaims "all Men . . . are endowed by their Creator."

But Scalia also believes that,

> The Declaration of Independence, however, is not a legal prescription conferring powers upon the courts. . . . . Consequently, while I would think it entirely compatible with the commitment to representative democracy set forth in the founding documents to argue, in legislative chambers or in electoral campaigns, that the state has no power to interfere with parents authority over the rearing of their children, I do not believe that the power which the Constitution confers upon me as a judge entitles me to deny legal effect to laws that (in my view) infringe upon what is (in my view) [an] unenumerated [unalienable] right.

One can quarrel with Justice Scalia's personal belief that parents have a God-given unalienable right to prevent grandparents from visiting their grandchildren, even when such visits are in the grandchildren's best interest. I would certainly want to know where such a right comes from and why grandparents do not have a countervailing right to at least some visitation. But it is difficult to quarrel with his legal conclusion that if the Constitution does not accord parents this exclusive right, the state retains the power to strike an appropriate balance, so long as it legislates in a reasonable manner.

When I was a child, I once asked my father why the mezuzah—the religious object that adorns the doorpost of a Jewish home—is always placed on a slant. My father asked our rabbi, who explained: "There were two schools of thought: One believed that it should be placed horizontally, the other vertically. Each was convinced it was correct but could not persuade the other. Finally, they split the difference by agreeing that it should be placed at a slant halfway between horizontal and vertical." What a wonderful symbol for a home, where compromise is always required. It is also a symbol for how America, at its best, has sometimes compromised and avoided the religious and political polemicism of other nations.

We should not strive for the uniformity of one absolutely correct morality, truth, or justice. The active and never-ending process of moralizing, truth searching, and justice seeking are far superior to the passive acceptance of one truth. The righting process, like the truthing process, is ongoing. Indeed, there are dangers implicit in accepting—and acting upon—any single philosophy of morality. Conflicting moralities serve as checks against the tyranny of singular truth. I would not want to live in a world in which Jeremy Bentham's or even John Stuart Mill's utilitarianism reigned supreme to the exclusion of all Kantian and neo-Kantian approaches;

nor would I want to live in an entirely Kantian world in which categorical imperatives were always slavishly followed. Bentham serves as a check on Kant and vice versa, just as religion serves as a check on science, science on religion, socialism on capitalism, capitalism on socialism. Rights serve as a check on democracy, and democracy serves as a check on rights.

Our constitutional system of checks and balances has an analogue in the marketplace of ideas. We have experienced the disasters produced by singular truths, whether religious, political, ideological, or economic. Those who believe they have discovered the ultimate truth tend to be less tolerant of dissent. As Hobbes put it: "Nothing ought to be regarded but the truth," and it "belongeth therefore to [the sovereign] to be judge" as to that truth. Put more colloquially, who needs differing—false—views when you have the one true view? Experience demonstrates we all do! The physicist Richard Feynman understood the lessons of human experience and the limitations of human knowledge far better than the philosopher Thomas Hobbes, as Feynman showed when he emphasized the basic "freedom to doubt"—a freedom that was born out of the "struggle against authority in the early days of science." That struggle persists.

Conclusion:
In the Meantime

As I complete this short book, the data keep changing: new variations of COVID are spreading; vaccines and treatments are improving; research is moving forward; change is the only constant. That is in the nature of all good science, especially with regard to illnesses with persistent variations. Despite the ever-changing scientific and medical data, several working conclusions, relevant to legal mandates, seem persuasive.

1. Current vaccines significantly reduce serious illness, hospitalization, and death among the vaccinated.
2. Widespread vaccination slows down the spread of COVID—and the seriousness of the illness among those who get it—to the vaccinated, and perhaps to the unvaccinated. It may also reduce the emergence of variants.
3. Receiving vaccinations is not entirely risk free for everyone in the short term, and we lack the data to be absolutely certain of long-term effects. Current data put the risk to healthy adults as extremely low. And for adults with certain medical conditions,

including allergies to components of available vaccines and problems related to their immune system, the risks of getting the vaccine must be balanced against the risks of not getting it.

4. Additional vaccinations, beyond the initial two, are now necessary for certain categories of adults and desirable for all others as the protective effects of initial vaccination dissipate over time. The possible advantages and disadvantages of additional vaccinations being from a different vaccine are being researched.

5. Vaccinating adolescents seems medically beneficial, and vaccinating younger children will likely prove beneficial as well.

6. It is unlikely that the various strains of COVID will entirely disappear in the short or middle term, as smallpox, measles, and polio did. It is likely to be more like the various strains of influenza that persistently recur with variations. As an article in the *Atlantic* was headlined: "The Coronavirus is here forever: this is how we live with it."[1] Citing experts, the article offered a somewhat optimistic, if cautious, prognosis:

> When enough people have gained some immunity through either vaccination or infections—preferably vaccination—the coronavirus will transition to what epidemiologists call "endemic." It won't be eliminated, but it won't upend our lives anymore.

With that blanket of initial immunity laid down, there will be fewer hospitalizations and fewer deaths from COVID- 19. Boosters can periodically reup immunity too. Cases may continue to rise and fall in this scenario, perhaps seasonally, but the worst outcomes will be avoided.

In light of the current science, which may change with emerging data, it is possible to draw implications for the future of how the law will deal with COVID mandates. As with good science, so too with good law: it must be sufficiently flexible to respond to changing circumstances, but it must also be sufficiently certain at any point in time to authorize necessary actions that are not universally accepted. The "meantime"—the time between the outbreak of a virus and the certain development of appropriate responses—is often when we must act. With these caveats in mind, let me first offer my personal judgment as to which actions should be authorized, and then my professional judgment as to what will be authorized.

I fervently hope the government will not have to compel vaccinations. Compulsion should generally be a last resort in a democracy committed to maximum individual autonomy. In several countries and areas of the United States, voluntary vaccination has achieved sufficiently high levels to avoid the need for compulsion. That would be the ideal situation, but it is unlikely to occur throughout the United States, where in some areas, and among some groups, the percentage of adults vaccinated is quite low. Because the virus and its variants quickly spread and are not limited to specific areas and groups, mandatory vaccination (with exceptions) may become necessary, if other steps—such as conditioning access to jobs and locations—prove insufficient to stem the contagion. The same is true of mandatory masking and distancing. Voluntary compliance is preferable but if compulsion is required to assure the safety of others, then compulsion in crowded places should be required to protect others. Certificates of vaccination should be available so the vaccinated can secure access to jobs and venues that legitimately require vaccination. Showing such a certificate is no greater infringement on liberty than showing identification for flying and entering buildings today. Even if the unvaccinated have the right to refuse vaccination, the vaccinated have the right not to take the risk of being in close proximity to them and to know whether they are vaccinated or not.[2]

If any compulsion is required, it should be explicitly authorized in a manner consistent with democratic governance: it should be voted by the legislature, signed by the executive, and upheld by the judiciary. Presidents, governors, mayors, and other executive officials should not be making controversial and intrusive laws. There may have been a time for emergency executive action, but that time has now passed. There may well be new emergencies that require immediate executive action, but legislatures are now on notice of the general issues raised by COVID. The time has come for legislative hearings, debate, nuanced deliberations, and voting. Not all the outcomes of such a democratic process will be sound. Moreover, there may be varying outcomes in different states and cities—which will be problematic for controlling a pandemic that, unlike the law, does not recognize political boundaries. That is the price we pay for our federal republic and our system of checks and balances.

The debate over these divisive and emotional issues should continue and there should be neither governmental nor private censorship of dissenting views.[3] False information should be quickly countered by truthful information so that the public can judge for itself based on the fullest information. Some people will make bad choices based on false information but that is also the price—and it may be an expensive one—we pay for freedom of speech and the right to dissent, which includes the right to be wrong.

Now for my professional analysis of how the courts will treat the changing legal responses to COVID and its inevitable variations. First, the courts, especially the Supreme Court, will try to avoid deciding the hardest ultimate issue: whether it is constitutional to compel unwilling adults to be vaccinated. The Justices will find reasons to decide cases on alternative grounds that do not require them to reach the ultimate issue. These alternative grounds include construing a statute to avoid a constitutional question, "standing," "lack of authority," or "insufficient exemptions." I will briefly describe these legal concepts.

The doctrine of "constitutional avoidance" was best articulated by Justice Louis Brandeis and developed by my teacher and mentor, Professor Alexander Beckel. Brandies listed a "series of rules under which [the Court] has avoided passing on a large part of all the constitutional questions pressed upon it for decision." Justice Felix Frankfurter described this doctrine of judicial restraint as the "one doctrine more deeply rooted than any other in the process of adjudication." Professor Beckel denominated this approach as "the passive virtues" and loved to quote Brandeis as saying, "The most important thing we do is not doing."[4]

"Not Doing" refers to not doing constitutional decision-making. In order to avoid making a constitutional decision, the courts often interpret ambiguous statutes in a manner that negates claims of unconstitutionality. For example, they read into the statute exemptions—such as medical, religious, or even ideological—that render the statute constitutional. Or they decide that the person challenging the constitutionality of the law is not directly affected by it. In the case of COVID, the courts may well rule that particular mandates may not properly be imposed by executives, rather than legislatures.

These and other techniques of avoidance have been used by courts, including the Supreme Court, in order to avoid constitutional issues. I think that process will continue for as long as possible. Ultimately, however, the Supreme Court may have to face the issue of compulsion. I believe it will uphold mandatory vaccination and other compelled compromises with liberty, if the medical science at the time demonstrates persuasively that compulsion is necessary to avoid serious public health consequences.

Just as the science adapts to the most current information, so, too, will the law adapt. The difference is that science is more impersonal than the law: it doesn't depend as much on the political and ideological leanings of the scientists; whereas some court decisions are influenced by the personnel on the court, their political and ideological leanings, as well as their approach to constitutional adjudication. Nevertheless, I predict that some forms of compulsion will likely be upheld by the current and future courts.

We live in an age of change, division and fear. All of these factors challenge the ability of our democracy to deal with controversial issues that have deep emotional components. I am confident, however, that our well-tested system of checks and balances will survive COVID, as will we as a brave and resiliant people.

Appendix A:
The Most Fundamental
Limitation on State Power

Appendix A: Because the philosophy of John Stuart Mill is so important to my arguments, I am appending the relevant portions of an essay I wrote about Mill.

The Most Fundamental
Limitation on State Power

Though I do not subscribe to the philosophy of any particular school of jurisprudence—I hope that I think for myself—I am a committed civil libertarian. The classic formulation of civil liberties was articulated by John Stuart Mill a century and a half ago. In 1993, Bantam Books decided to publish a new edition of Mill's philosophy, On Liberty and Utilitarianism, *and asked me to write an introductory essay on its influence. In that essay, I try to apply Mill's thinking to current issues and to suggest some critiques.*

The Principle

Few principles of civil morality have had so profound an intellectual influence within Western democracies as John Stuart Mill's "one very simple principle." The principle, governing the proper allocation of state power and individual liberty, was articulated by Mill in his 1859 essay entitled "On Liberty." In Mill's own words:

> That principle is, that the sole end for which mankind are warranted, individually or collectively, in interfering with the liberty of action of any of their number, is self-protection. That the only purpose for which power can be rightfully exercised over any member of a civilized community, against his will, is to prevent harm to others. His own good, either physical or moral, is not a sufficient warrant. He cannot rightfully be compelled to do or forbear because it will be better for him to do so, because it will make him happier, because, in the opinions of others, to do so would be wise, or even right. These are good reasons for remonstrating with him, or reasoning with him, or persuading him, or entreating him, but not for compelling him, or visiting him with any evil in case he do otherwise. To justify that, the conduct from which it is desired to deter him, must be calculated to produce evil to some one else. The only part of the conduct of any one, for which he is amenable to society, is that which concerns others. In the part which merely concerns himself, his independence is,

of right, absolute. Over himself, over his own body
and mind, the individual is sovereign.

Mill made it clear that his principle applied only to "human
beings in the maturity of their faculties" and granted to the
state the power to determine, within reason, the age "of
manhood and womanhood." The explicit inclusion of womanhood reflected more than syntactical completeness; Mill
wrote eloquently in favor of women's equality in the home,
at the ballot box, and in the world at large.

While support for women's rights was uncharacteristic of
his circle during the mid-nineteenth century, Mill's implicit
acceptance of colonialism was all too typical. He exempted
from his principle "those backward states of society in which
the race itself may be considered as in its nonage." For such
"barbarians," Mill paternalistically concluded, benevolent
"despotism is a legitimate form of government," since liberty
has no application "to any state of things anterior to the
time when mankind may have become capable of being
improved by free and equal discussion."

But neither his progressive inclusion of women nor his
regressive exclusion of "backward" people is central to Mill's
principle and its remarkable influence on Western society.
Like other profoundly influential principles, such as the
Bible's "Thou shalt love thy neighbor like thyself" and Kant's
"So act, that the rule on which thou actest would admit of
being adopted as a law by all rational beings," the principle
itself is as simple as it is eloquent (at least in conception—
Mill was the first to acknowledge its difficulties in application, leaving that to a sketchy final chapter that is among

the weakest in an otherwise persuasive essay). The power of the state may not be used to compel a reasoning adult to do or not do anything solely because such action or inaction would be better for that adult.

It is interesting that this principle was, for Mill, based entirely on utilitarian considerations: "It is proper to state that I forgo any advantage which could be derived to my argument from the idea of abstract right [since] I regard utility as the ultimate appeal on all ethical questions." There are, however, persuasive utilitarian arguments in favor of compelling adults to do certain things that would make them happier and better people. Indeed, if a truly benevolent despot really knew the secret of maximizing happiness for everyone, there would surely be many utilitarians who would feel compelled to grant him the power to do what no democracy has thus far succeeded in doing: namely, producing a universally happy society.

In the end, however, Mill is not at his best in attempting to justify his principle solely on conventional utilitarian grounds. Though Mill himself eschews all advantage to his argument from "abstract right," that does not necessarily mean that those who reject utilitarianism and accept abstract rights must reject Mill's principle. Even as an abstract right or as part of a rights-based system, Mill's principle has much to commend it. This is an instance where the power of the principle transcends the strength of the underlying justification offered by its proponent. I think it is true today that a considerable number of non-utilitarians do, in fact, accept Mill's basic principle with as few or as many variations as orthodox utilitarians who accept it.

Indeed, it is fair to say that the fundamentals of Mill's principle have become almost a conventional wisdom of Western society, at least among its intellectuals. It is generally taken for granted as a premise of debate concerning the proper allocation of state power and individual freedom. To be sure, there are some state paternalists, especially among the religious ultraright, who still believe that it is the proper function of government to compel adults to do what is deemed best for them. But the vast majority of contemporary Western thinkers—whatever their philosophical bent—seem to accept the basic Millian principle that it is not the proper function of government to compel conduct solely in order to improve the life of an adult who does not necessarily want his or her life so improved.

Many philosophers reject the rigidity with which Mill stated his thesis. Others have greater difficulty than even he had in clearly distinguishing between actions that affect only the actor and those that have a discernible impact on others. But it is not easy to find many who categorically reject the core concept central to Mill's principle and who would grant the state the power to make reasoning adults take nontrivial actions that they have knowingly chosen not to take but that the state believes they should take in order to better themselves or make them happier. This is especially true in a nation as diverse and heterogenous as the United States, where it would be difficult to reach a consensus on what constitutes the kind of betterment of happiness that could properly be imposed. But even in more homogenous democratic nations, Mill's core principle has become the conventional wisdom, at least in theory.

The best evidence of how influential Mill's principle has become—indeed, how it is presumed by most thinkers—may be the repeated efforts of those who would compel a given action against protesting individuals to rationalize such force by reference to the rights of others rather than by reference to the good of the compelled individual. Examples abound, but one will suffice to make the general point. A distinguished colleague of mine would seek to justify mandatory seat belt laws by rejecting the argument that "only the belt-wearer's own welfare [is] at risk." He argues instead that we should recognize that

> refusing to buckle up endangers innocent third parties, not only the dependent children of those who insist on not buckling, and not only those who end up paying higher insurance premiums and higher taxes so that others may enjoy the "freedom" not to buckle, but also those who end up being injured or even killed in avoidable collisions when unbuckled drivers lose control of their cars. Quite simply, the seat-belt law prevents people from becoming loose objects when a car skids or veers into a tree or another vehicle; a belted driver is less likely to become a helpless spectator as his car is turned into an unguided missile. Surely that is a legitimate exercise of society's power to protect the innocent, not the entering wedge of tyranny.

While these observations may all have some small validity, they miss the big picture, namely, that seat belt laws have as their primary object the mandatory protection of the adult belt wearer. I, too, favor mandatory seat belt laws, but I recognize that support for such paternalistic legislation requires a compromise with Mill's principle. And it is a compromise I am prepared to make explicitly rather than uncomfortably to try to squeeze seat belt laws into Mill's principle by invoking flying people and leaping logic.

My compromise would establish two significant exceptions to Mill's principle. The first I call the "light pinky of the law" exception. The second I call the "Thanks, I needed that" exception.

The "light pinky of the law" is at the opposite end of the continuum from the "heavy thumb of the law." It refers to regulations carrying minor financial penalties that are calculated to influence the behavior of people who really have no ideological objection to doing something that will help them, but who don't care enough to take the step without some gentle nudging force from the law. Seat belt laws are a perfect example. Most Americans will wear seat belts if the law requires them to and many will not wear them if the law does not require them to. That may seem silly to any believer in rational, cost-benefit analysis. Why, after all, should a fifty-dollar fine work when the compelling statistical and clinical evidence that safety belts save lives does not work? The answer lies in the indisputable fact that most people do not rationally calculate the costs and benefits of their actions, particularly when the benefit is hypothetical, long-term, and statistically quite unlikely to come about. That is so even if the cost is as trivial as buckling up.

For a variety of reasons, the law often works where rational calculation does not. People do not generally want to be perceived—by themselves or others—as lawbreakers even when the penalty is quite trivial. The law does have some kind of moral imperative that moves people to action and inaction more powerfully than the mere economic cost attached to violation. To be sure, if the law is overused, or is used immorally or foolishly, much of that moral imperative may be diluted. But as of now, for most citizens of Western democracies, the law does work, at least in situations where it is used to nudge people into doing something relatively cost-free that promises some potential benefit.

That is why I favor mandatory seat belt laws and other simple self-helping safety rules that are enforced with no more than small fines. But the "light pinky of the law" exception to Mill's principle should not, in my view, be expended beyond the narrow areas in which it is appropriate. To make my point, I will argue that mandatory motorcycle helmet laws—though similar in many respects to seat belt laws—may exceed the narrow bounds of my exception. The distinction may be subtle, but it is real: Many car drivers who would not wear seat belts if the law were silent are not conscientiously opposed either to seat belts or to the legal requirement that they be worn; they are simply lazy, forgetful, or unconcerned; they will do whatever the law nudges them to do. Many motorcycle riders who would not wear helmets in the absence of a law seem to be conscientiously opposed both to helmets and to the legal requirement that they wear them. If I am right about that difference, then mandatory helmet laws are really different from mandatory seat belt laws—at least for those cyclists who care deeply about their freedom to maim and kill themselves. For the conscientiously opposed cyclist—as distinguished from the car driver who couldn't care less whether he buckles up or doesn't—the legal requirement that he wear a helmet will be perceived as a fundamental denial of freedom rather than as a trivial nudge from the state. He will feel the heavy thumb of the law upon him rather than the light pinky that will be felt by the typical car driver who would not buckle up if he did not "have to."

But what about those few car drivers who feel as strongly about seat belt laws as the helmet-free cycle fanatics feel about the helmet laws? There are two ways of dealing with this minority: If we lived in a totally honest society where all defendants always told the truth about why they violated the law, there could be an exception written into the seat belt law for conscientious objectors who could show that they had thought through the issue and had come to an ideological position against buckling up (or against being compelled to buckle up). But because many people who were caught unbuckled would falsely claim that they were conscientious objectors when they were merely lazy, the exception might swallow up the rule. The other way of dealing with the small number of conscientious objectors is simply to regard the fifty-dollar fine as a tax or an insurance surcharge for engaging in behavior that is dangerous to themselves but for which society in general will have to pay. In other words, society would be telling these people that they are not forbidden from driving unbuckled; they must simply pay a small price for doing so.

In no case, under the "light pinky of the law" exception, would I ever put a dissenter in prison—or punish him or her harshly—for refusing to take an action that benefits only him or her. I would reserve serious penalties for those who squarely fit within Mill's principle.

This brings us to the second exception, which, in my view, sometimes justifies mandatory seat belt laws designed to prevent injuries to those who would not otherwise buckle up, as well as some other limited state force designed to help only the compelled individual. The "Thanks, I needed that" exception derives from the typical scene in old grade-B movies in which one character is out of control and the other character slaps him in the face to restore his control. The slapped character invariably says, "Thanks, I needed that," thus demonstrating his after-the-fact appreciation of his friend's paternalistic assault. Even Mill would permit state compulsion to prevent the mentally ill—those not capable of rational thought—from harming themselves.

But my exception would, perhaps, go a bit further. I would justify state compulsion to prevent—at least temporarily—a distraught but rational adult from killing (or otherwise inflicting irreversible serious harm on) himself or herself. I would regard it as morally permissible—indeed, perhaps morally imperative—to try and prevent such self-inflicted harm if I could do so without unreasonable risk to myself or others. I would do so in the expectation that after the person calmed down and thought it through, he would thank me—perhaps not literally but at least in his own mind. If I were wrong in a particular case, I would still not regret what I had done, because the person has an eternity to be dead, and I would not regard myself as having denied him much if I deprived him of several additional hours or even days of death. If, on the other hand, I were to err on the side of not preventing the suicide of a person who would indeed have thanked me for doing so, then I would have contributed to denying him the rest of his life.

As with the motorcycle helmet example, I would not apply the thank-you exception to rational adults who have carefully thought through the issue of suicide over a substantial period of time and have decided to end their lives.

It is somewhat more questionable whether seat belt laws fit comfortably within the thank-you exception as well. The vast majority of car drivers who grumble over buckling up would certainly say thank you if they were involved in an accident in which their lives (or limbs) were saved by wearing the seat belt they would not have worn but for the law. But would they say thank you after each car trip during which they were required to buckle up, or only when—and if—they were involved in an accident?

There is a considerable danger in expanding the thank-you exception to a point where it could swallow up much of Mill's basic principle. A large number of hypothetical paternalistic compulsions—for example, those directed against smoking, overeating, or not exercising—could be justified by reference to a mirror-image version of the thank-you exception. I can easily imagine angry people on their deathbeds complaining about the lack of compulsion that allowed them to smoke, eat, and couch potato themselves to death. "Why didn't you make me stop smoking? I would be thanking you today if you had!" Well, one response to that hypothetical conversation is: "No, you wouldn't be thanking me if you were up and around and healthy, because you wouldn't appreciate—as you now do—the importance of not smoking. It required you to come face-to-face with death for you to understand why you should not have smoked, and now it is too late." The more persuasive answer is that there is a crucial difference between a brief one-short act of compulsion such as preventing the distraught person from jumping out the window or taking poison and a long-term, lifestyle-changing compulsion such as that required to make a person stop smoking, overeating, or not exercising. The state should be far more reticent about enforcing long-term, lifestyle-changing compulsions on unwilling adults than it should be to risk not being thanked for a brief one-shot interference with an adult's liberty that may well be appreciated in retrospect.

I offer these two limited exceptions to Mill's principle to suggest that it is far better to argue about the limits of the principle itself than to accept it as an almost biblical (or constitutional) rule of action and then try to find ways to squeeze what are really exceptions into the parameters of the principle.

We live today in a far more interdependent society than the one in which Mill lived. Even in Mill's time and before, there were those who believed that "no man is an island, entire of itself." Mill recognized, of course, that actions that cause harm to the actor often create ripples that touch others. As we shall see later, however, Mill is not at his best in dealing with such matters of degree. Nor is it clear how Mill would have applied his principle to somewhat more complex and multifaceted problems than those he discussed.

Consider, for example, some current controversies on which Mill's principle may bear differently in today's America than it appeared to bear in Mill's England. Mill may or may not have known that smoking harms the lungs and heart of the smoker. But even had he known that fact, he would still not countenance legislation banning smoking. He might, perhaps, have approved of labeling laws designed to give the smoker information necessary to decide whether the present pleasure of the puff was worth the possible pain of the future. Today we know that smoking hurts not only the lungs and hearts of smokers but also the health of nonsmokers. That might well have led Mill to conclude that adults have the right to inhale but not to exhale—at least not in the presence of nonconsenting adults or children. Just as your right to swing your fist ends at the tip of my nose, so, too, your right to puff on a cigarette ends at the edge of my nostrils.

In Mill's day—indeed until quite recently—pornography and obscenity were regarded as "moral" issues akin to masturbation. Both were thought to be bad for the soul, the psyche, and the sexuality of the viewer or reader. As such, Mill would find no basis for preventing adults from indulging in smut in the privacy of their bedrooms. Now, however, we are told by some feminists that those who view or read pornography will be more likely to engage in violent actions against nonconsenting women. This is not the place to rehearse the empirical debate over whether pornography causes rape or other violence toward women. The issue here is a normative one: if it could be shown that pornography did cause harm not only to its consumers but also to others who do not consent to its availability, may the state properly prevent its consumption, even in private?

A similar controversy, but with an interesting twist, surrounds the state regulation of addictive drugs, especially heroin. By criminalizing heroin—a chemical that harms the user but does not itself make him or her more prone to violence—the state increases the cost of obtaining the highly addictive drug. The "market" cost of heroin would be quite low if it were available by medical prescription, but because it is illegal, its cost is many times higher. This increased cost causes most heroin addicts to commit many more acquisitive and predatory crimes against innocent people than they might otherwise commit. (I say "otherwise," because many heroin addicts have long criminal backgrounds.) Accordingly, the criminalization of heroin violates Mill's principle in two ways: first, it employs the power of society to compel (or at least try to compel) the adult user to forbear from doing something because not doing it would be better for him or her; second, by doing so, it causes harm to others.

This may sound like a simplistic analysis, since the causes of crime and the effects of addiction are so complex and varied. Moreover, this analysis is not as clearly applicable to other drugs, such as crack cocaine, which may themselves make the user more prone to violence. But the heroin example makes an important point about the misuses of the criminal sanction.

Mill spoke indirectly to this issue in the context of prostitution and gambling. He concluded that "fornication" and "gambling" must be tolerated, but then he asked whether a person should "be free to be a pimp, or to keep a gambling-house?" He would probably have come to the same conclusion and asked the same question about the drug user (at least those who retain the power of rational thought) and the drug seller. Mill regarded the question of such professional accessories as "one of these which lie on the exact boundary line." It was clear to him, as it remains clear today to most civil libertarians—though not to all feminists—that the case for criminalizing the professional purveyor of vice is far stronger than the case for criminalizing the occasional consumer of vice, despite the reality that without consumers there would be no market.

Another controversial set of contemporary issues also demonstrates the limitations of Mill's principle. The whole area of "fetal" rights is not really amenable to solution by reference to Mill's principle because the essential dispute is over a question that Mill did not address; namely, is the fetus a part of the carrying woman and thus beyond the ken of compulsory state regulation? Or is the fetus a "someone else" that the state has the legitimate power to protect against abortion, abuse, or neglect?

Some argue, as the courts have sometimes implied, that the fetus becomes a "someone else" at the moment of viability—that is, when it would be capable of independent life outside the womb. Others argue that the fetus becomes a "someone else" when the carrying woman makes the decision to carry to term rather than to abort. Under this latter approach, the state might have the power to compel a pregnant woman who had decided not to abort to refrain from excessive drinking or other activities that pose significant health risks to the "someone else" she has decided to carry to term.

In the last analysis, Mill's principle does not help us decide whether or when a fetus becomes a "someone else"— that is for theologians, biologists, judges, or perhaps each pregnant woman to decide. But Mill's rule can help us sort through some complex philosophical issues regarding the relationships between carrying woman and fetus—once it is decided that the fetus has become a "someone else" deserving of some degree of state protection. A wise state may, of course, decline to exercise power—particularly the power of the criminal law—in certain areas where it may well have legitimate authority to act. The relationship between woman and fetus may be such an area.

One more general issue of complexity, alluded to earlier, may warrant brief further discussion. We live in an age in which people have become far more economically interdependent because of insurance, welfare, taxation, and other mechanisms for sharing the risks and costs of individual hardships. Thus, if some drivers buckle up and others do not, and if the cost of insurance or medical care will rise for all as the result of avoidable injuries caused by a driver's decision not to buckle up, then it can be argued that we all have a stake in every driver's buckling up. That argument can be taken, however, to absurd extremes. We non-skiers, non-bungee jumpers, non-hang gliders, also have a stake in preventing daredevils from taking what we regard as undue risks to their limbs and our pocketbooks. We exercising, cholesterol-watching, fat-avoiding, one-drink-a-day consumers have a stake in every greasy hamburger and kielbasa eaten by a couch potato whose clogged arteries will cost us money. Where would a reasonable line be drawn between compelling everyone to live a safe, healthy, moderate life and permitting undue-risk takers to have their destructive lifestyles (and death wishes) subsidized by the rest of us?

One way of dealing with this issue is to impose risk costs on certain clearly dangerous activities. We already do that through differential insurance premiums based on risk factors such as smoking and hang gliding. It would not be unreasonable, in states that make the wearing of seat belts optional, for insurance companies to give drivers who agree to buckle up a discount on the premium. Indeed, the state might even go further, in my view, and impose a tax on those who refuse to wear seat belts or motorcycle helmets. There already are special taxes in many states on cigarettes, the proceeds from which are used to reduce the societal costs attributable to smoking. I doubt that Mill would have had difficulty with a system that imposed the costs of risk taking more directly on the risk takers, so long as the risk taker remained free of state compulsion and could decide for him- or herself whether to incur the risk and the cost.

Mill recognized, of course, the interdependent nature even of his society. Paraphrasing John Donne, Mill wrote:

No person is an entirely isolated being; it is impossible for a person to do anything seriously or permanently hurtful to himself, without mischief reaching at least to his near connexions, and often far beyond them. If he injures his property, he does harm to those who directly or indirectly derived support from it, and usually diminishes, by a greater or less amount, the general resources of the community. If he deteriorates his bodily or mental faculties, he not only brings evil upon all who depended on him for any portion of their happiness, but disqualifies himself for rendering the services which he owes to his fellow creatures generally; perhaps becomes a burthen on their affection or benevolence; and if such conduct were very frequent, hardly any offense that is committed would detract more from the general sum of good. Finally, if by his vices or follies a person does no direct harm to others, he is nevertheless (it may be said) injurious by his example; and ought to be compelled to control himself, for the sake of those whom the sight or knowledge of his conduct might corrupt or mislead.

Having recognized this interdependence, Mill proceeded to reject it on relatively unsophisticated grounds:

But with regard to the merely contingent, or, as it may be called, constructive injury which a person causes to society, by conduct which neither violates any specific duty to the public, nor occasions perceptible hurt to any assignable individual except himself; the inconvenience is one which society can afford to bear, for the sake of the greater good of human freedom. If grown persons are to be punished for not taking proper care of themselves, I would rather it were for their own sake, than under pretense of preventing them from impairing their capacity of rendering to society benefits which society does not pretend it has a right to exact.

In the end, Mill merely gives us his preference: "the greater good of human freedom" over what he calls "the inconvenience" of "constructive injury" caused by the exercise of that freedom. It is a preference shared by most libertarians and individualists but not one shared by all communitarians or even utilitarians. This conflict has divided and will continue to divide people of goodwill who care about both freedom and responsibility.

Endnotes

Introduction

1. The abortion issue is more complex than the gay rights issue as I wrote back in 2004:

   "Some arrogantly proclaim that if people only thought more clearly and morally, they would all understand that abortions violate the natural right of the living fetus. Others argue that if people thought more logically, they would all understand why compelled birth violates the natural rights of the mother. Both are demonstrably wrong, since clear-thinking, moral people come to diametrically opposite conclusions based on their differing worldviews, values, upbringings, experiences, and assessments of the future. Each side may well be right, if one accepts the premises of its arguments. Some of the premises are plainly empirical; for example, easy abortion promotes sexual promiscuity. Others are plainly faith-based; for example, God has ordained that no abortions—or only some abortions—may be performed. Still other views purport to be based entirely on moral claims without regard to facts—but upon analysis, they turn out to have a significant empirical component; for example, if abortion is permitted, it will diminish the value we place on human life and make it easier for us to justify other takings of human life. His sort of slippery-slope argument is premised on empirical claims that may be difficult to establish, but they are empirical claims, nonetheless.

For many, experience is the determining factor in the abortion debate. They look to the experience of countries that have prohibited all abortion and they see young women who have been killed or injured by illegal abortionists performing surgery in back rooms without proper medical care. They see the hypocrisy of the rich being able to secure medical abortions in other countries while the poor are denied that option. They see the difference between preserving the life of an eighteen-year-old sentient woman and preserving the opportunity for a month-old fetus to be born." *Rights from Wrongs* (2004)

There are no pervasive countervailing arguments for discriminating against gay people in their desire to marry. No third entity, like a fetus, is involved, and it is nobody's business—certainly not the state's—to determine whether a competent adult may have consensual sex with another competent adult of the same sex.

2. For a more complete analysis of my views regarding Mill see Appendix A.

3. Mandatory seatbelt laws pose a challenge to Mills' doctrine. Here is what I have written about that issue: "I favor mandatory seat belts and other simple self-helping safety rules that are enforced with no more than fines. But the "light pinky of the law" exceptions to Mill's principle should not, in my view, be expended beyond the narrow areas in which it is appropriate." My views are spelled out more completely in Appendix A.

4. See discussion of human papillomavirus (HPV) vaccine in "5 Reasons Boys and Young Men need the HPV Vaccine too" in MSKcc.org, Jun 10, 2021. Some research suggests that male circumcision may reduce certain health risks for female sex partners. But the research is too uncertain to warrant compulsion.

5. See Tversky and Khanaman, *Judgment under Uncertainty: Heuristics and Biases* (1974); see also Neil Kakkar, *Bayes Theorem: A Critical Framework for Critical Thinking*, Aug. 4, 2020.

6. See Alan Dershowitz, *Preemption: A Knife that Cuts Both Ways* (2006).

7. Some extremists have even taken the position that voluntary vaccination should be banned by federal authority because the vaccine is unproven and may cause harm. I will not be including those and other "conspiratorial" views in my discussion.

8. See Chapter 2.

9. See Brent Kendall, "Supreme Court Rejects Request to Block Indiana University's Vaccine Mandate for Students", *Wall Street Journal*, August 12, 2021.

10. See Chapter 4.

11. See Jacobson v. Massachusetts, 197 U.S. 11 (1905).

12. See Fn 6, supra.

13. See Alan Dershowitz, "Could It Happen Here: Civil Liberties and a National Emergency," *Nation*, March 15, 1971; also *Shouting Fire*, 2002

14. See John Yoo & Robert Delahunty, "Biden's Eviction Moratorium Extension is Executive Overreach," *National Review*, August 11, 2021.

15. See Brent Kendall, "Supreme Court Lifts Part of New York's Eviction Moratorium," *Wall Street Journal*, August 12, 2021.

16. See fn. 10.

17. See Lisa Lernet, "How Republican Vaccine Opposition Got to this Point," *New York Times*, July 17, 2021.

18. NPR, Dec. 20, 2020, "Race and the Roots of Vaccine Skepticism."

19. See Chapter 8.

20. See Chapter 2.

21. My emails and Twitter responses are filled with ad hominem attacks and threats whenever I write or speak about the pandemic.

22. "We must all hang together or most assuredly, we will all hang separately."

Chapter 1: The Case for Compulsion: From Easy to Hard

1. Mill and Kant would give different answers to the above questions.

2. *District of Columbia v. Heller*, 554 U.S. 570 (2008).

3. Those who claim a health problem have the option of staying away from places that require masks or seeking an exception.

4. This sentence would seem relevant to those who find wearing a mask "destressing, inconvenient, or objectionable."

5. "Mass, successful vaccination programs have meant deadly contagious viruses and diseases such as polio, tuberculosis and measles have been largely eradicated in parts of the world or greatly suppressed by vaccination programs and the herd immunity they foster." Holly Ellyatt, Here's Why Herd Immunity from Covid is "Mythical" with the Delta Variant, Aug. 12, 2021.

6. Ibid.

7. Ibid.

8. Ibid.

9. Ibid.

10. Ibid.

11. Daniel P. Oran, Eric Topol, "The Crucial Vaccine Benefit We're Not Talking About Enough," *Scientific American*, July 27, 2021.

12. Laure Wasmley, "Vaccinated People with Breakthrough Infections can spread the Delta Variant, CDC says," *NOR*, July 30, 2021.

13. *New York Times*, Aug. 19. 2021.

14. When the decision was overruled several decades later, a cynic observed, referring to the generations of judges that had implicitly upheld that benighted decision, that "Three generations of imbeciles are enough."

Chapter 5: Debating Vaccination with Robert F. Kennedy Jr.

1. https://childrenshealthdefense.org/transcripts/robert-f-kennedy -jr-vs-alan-dershowitz-the-great-vaccine-debate/.

2. https://www.youtube.com/watch?v=IfnJi7yLKgE.

Conclusion

1. Sarah Zhang, *Atlantic*, August 22, 2021.

2. Medical privacy raises difficult issues. For example, Stanford University has refused to tell students whether their assigned roommates are vaccinated.

3. The issue of doctors conveying false medical information raises different and more complex issues.

4. Andrew Nolan, *The Doctrine of Constitutional Avoidance: A Legal Overview*, Congressional Research Service, Sept. 2, 2014.

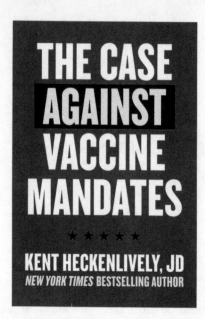

For the opposite argument, please read *The Case Against Vaccine Mandates* by Kent Heckenlively.

Kent Heckenlively, *New York Times* bestselling author of *Plague of Corruption*, calls upon both common sense and legal precedence to fight against vaccine mandates around the country.

"My body, my choice!" used to be the rallying cry of the left in the abortion fight. But now this same principle of bodily autonomy is the central argument of conservatives, such as that of Florida Governor Ron DeSantis in fierce opposition to so-called "vaccine passports," which would limit whether an individual could attend movies or other public events, work, or even go to school, if they chose to decline a COVID-19 vaccine.

While cities like New York close their doors to unvaccinated people, the fight against vaccine mandates is cobbling together an unexpected alliance across the political spectrum, such as the Black mayor of Boston, Kim Janey, who recently claimed, "there's a long history" in this country of people "needing to show their papers" and declaring any such passport as akin to slavery.

The starting point agreed upon by all parties as to whether the government can bring such pressure to bear upon individuals is the 1905 US Supreme Court of *Jacobson v. Massachusetts*. In that case, a Lutheran pastor declined a smallpox vaccination and was fined $5, the equivalent of a little more than $150 in today's currency, or less than many traffic tickets. The *Jacobson* case sparked a shameful legacy in American jurisprudence, being used as the sole reasoning by the US Supreme Court to allow the forced sterilization of a female psychiatric patient in 1927. This ruling paved the way for the involuntary sterilization of more than sixty thousand mental patients and gave legal justification to the eugenics movement, one of the darkest chapters in American medicine.

In *The Case Against Vaccine Mandates*, *New York Times* bestselling author Kent Heckenlively, whose books have courageously taken on Big Pharma, Google, and Facebook, now points his razor sharp legal and literary skills against vaccine passports and mandates, which he believes to be the defining issue as to whether we continue to exist as a free and independent people.